Power Maths

Year 3 Textbook

Series Editor: Tony Staneff

Dexter

Dexter is determined.

He tries hard and never gives up.

brave

Astrid

curious

Ash

flexible

Flo

helpful

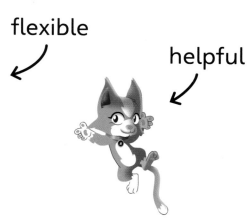

Sparks

Pearson

Contents

This tells you which page you need.

I cannot wait to start!

How to use this book

These pages make sure we're ready for the unit ahead. Find out what we'll be learning and brush up on your skills.

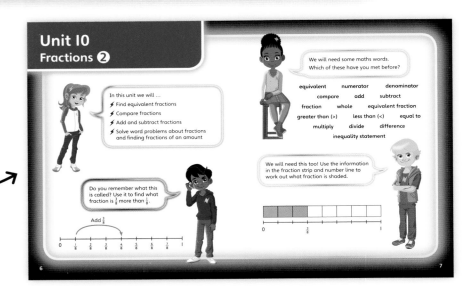

Discover

Lessons start with **Discover**.

Here, we explore new maths problems.

Can you work out how to find the answer?

Don't be afraid to make mistakes. Learn from them and try again!

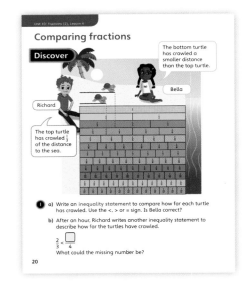

Share

Next, we share our ideas with the class.

Did we all solve the problems the same way?
What ideas can you try?

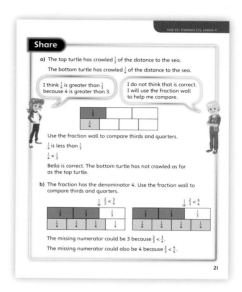

Think together

Then we have a go at some more problems together. Use what you have just learnt to help you.

We'll try a challenge too!

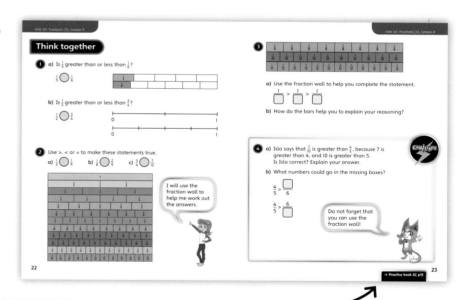

This tells you which page to go to in your **Practice Book**.

At the end of each unit there's an **End of unit check**. This is our chance to show how much we have learnt.

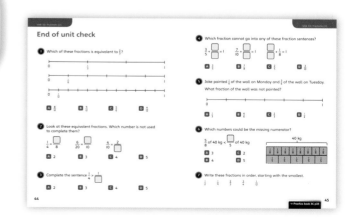

Unit 10
Fractions ②

In this unit we will ...

⚡ Find equivalent fractions

⚡ Compare fractions

⚡ Add and subtract fractions

⚡ Solve word problems about fractions and finding fractions of an amount

Do you remember what this is called? Use it to find what fraction is $\frac{3}{8}$ more than $\frac{1}{8}$.

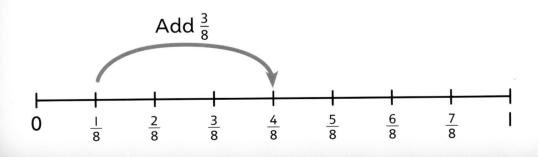

Add $\frac{3}{8}$

$$0 \quad \frac{1}{8} \quad \frac{2}{8} \quad \frac{3}{8} \quad \frac{4}{8} \quad \frac{5}{8} \quad \frac{6}{8} \quad \frac{7}{8} \quad 1$$

We will need some maths words.
Which of these have you met before?

equivalent   numerator   denominator

compare   add   subtract

fraction   whole   equivalent fraction

greater than (>)   less than (<)   equal to

multiply   divide   difference

inequality statement

We will need this too! Use the information in the fraction strip and number line to work out what fraction is shaded.

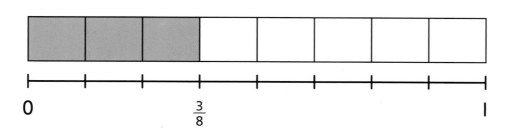

0    $\frac{3}{8}$ 1

Equivalent fractions 1

Discover

1 **a)** Who is correct, Lee or Mr Lopez?

b) Look at the lines drawn on the track.

Write two or more fractions that are equal to $\frac{1}{2}$.

Share

a)

> This model represents $\frac{1}{2}$ of the whole journey. There are 2 parts in the whole and 1 is shaded.

Look at the fractions $\frac{1}{2}$ and $\frac{2}{4}$.

They have different numerators and denominators, but show the same distance.

Both Lee and Mr Lopez are correct.

> Each part is cut into 2 equal parts. Lee has run $\frac{2}{4}$ of the journey. $\frac{1}{2} = \frac{2}{4}$ so these are **equivalent fractions**.

b)

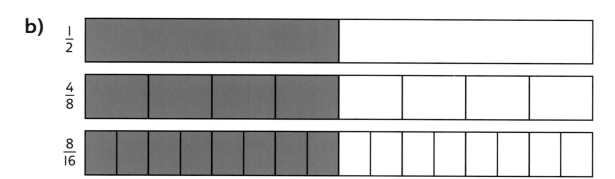

$\frac{1}{2} = \frac{2}{4}$ $\qquad$ $\frac{1}{2} = \frac{4}{8}$ $\qquad$ $\frac{1}{2} = \frac{8}{16}$

These are all equivalent fractions.

> I folded a strip of paper to help me find the different fractions.

Think together

1 Lexi folds a paper strip into 3 equal parts.

She colours 1 of the parts.

She folds the strip in half, across the length, then unfolds it.

a) What fraction of the strip is coloured?

b) Write an equivalent fraction for this.

$$\frac{\boxed{}}{3} = \frac{\boxed{}}{6}$$

2 Jamilla has a different paper strip.

She folds the strip into 5 equal parts.

She colours 1 part.

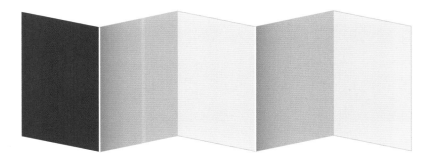

I am going to do the same as Jamilla. Then I will fold the paper in half to see what happens.

She folds the strip in half, across the length.

a) What fraction of the strip is coloured?

b) Write an equivalent fraction for this.

$$\frac{\boxed{}}{\boxed{}} = \frac{\boxed{}}{\boxed{}}$$

3 Each fraction is represented by a colour.
Write the missing fractions.

CHALLENGE

1

$\frac{1}{2}$

What equivalent
fractions can you write?

I think there is more than one answer
for some of these. I will write all the
equivalent fractions I can see.

→ Practice book 3C p6

Equivalent fractions ②

Discover

Mo

Kate

Kate's bracelet

1 **a)** $\frac{3}{4}$ of Kate's bracelet is blue.

What other fraction on the fraction wall is equivalent to $\frac{3}{4}$?

b) Find another pair of equivalent fractions on the fraction wall.

Are there any more equivalent fractions?

Share

a)

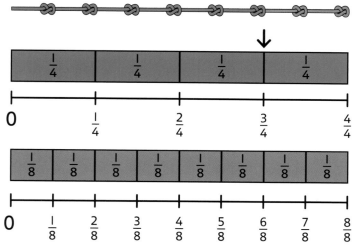

I can show fractions on a number line.

$\frac{3}{4}$ and $\frac{6}{8}$ are equal: $\frac{3}{4} = \frac{6}{8}$. These fractions are equivalent.

b)

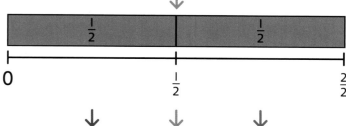

I looked down the number wall to see which fractions lined up.

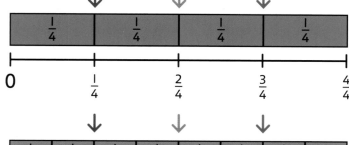

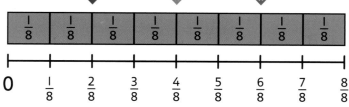

The other equivalent fractions on the fraction wall are:

$\frac{1}{4} = \frac{2}{8}$ $\frac{1}{2} = \frac{2}{4} = \frac{4}{8}$ $\frac{3}{4} = \frac{6}{8}$

The number line and fraction wall can help us see when two fractions are equal.

13

Think together

1 Complete the missing fractions on the number lines.

a)

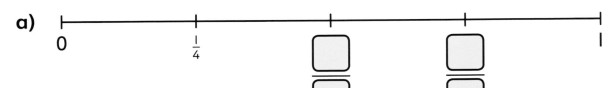

b)

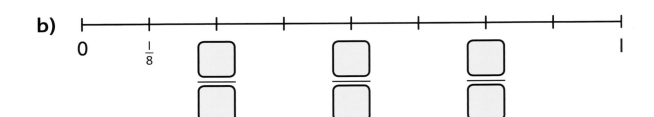

c)

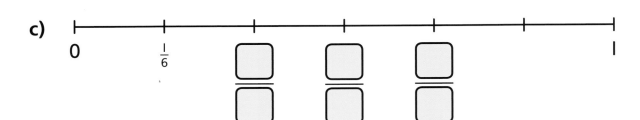

d)

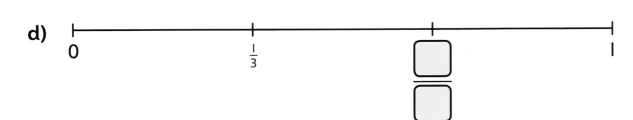

2 Look at the number lines from question 1. Use them to work out the equivalent fractions.

a) $\frac{1}{4} = \dfrac{\square}{\square}$

c) $\frac{2}{3} = \dfrac{\square}{\square}$

e) $\frac{2}{4} = \dfrac{\square}{\square} = \dfrac{\square}{\square}$

b) $\frac{1}{3} = \dfrac{\square}{\square}$

d) $\frac{3}{4} = \dfrac{\square}{\square}$

3 Find fractions that are equal in size to $\frac{2}{3}$.

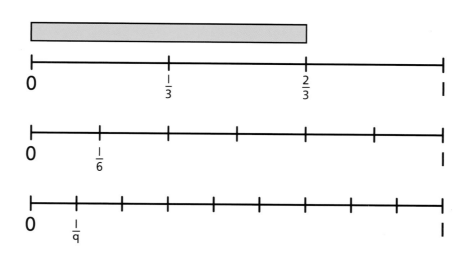

I am going to look for fractions that are an equal distance from 0 on the number line.

4 Zac thinks that $\frac{3}{4} = \frac{7}{8} = \frac{11}{12}$.

CHALLENGE

a) Explain why Zac is wrong.

b) Find two fractions that are not equal to $\frac{3}{4}$.

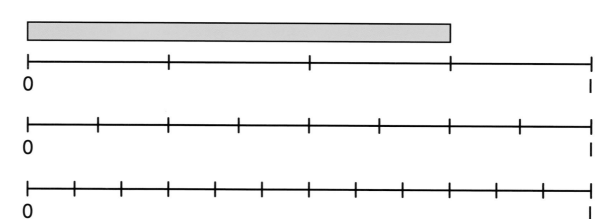

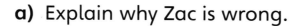

I am going to look for fractions on the number line that are different lengths from 0.

15

→ Practice book 3C p9

Equivalent fractions ③

Discover

Reena

Danny

I will make a fraction that is equivalent to $\frac{1}{2}$.

1 a) How can Reena use her remaining cards to complete the puzzle?

b) Danny uses four of his cards to make two other equivalent fractions.

How could Danny complete his puzzle?

Share

a) Use a fraction wall to check for equivalent fractions.

$\frac{1}{6}$	$\frac{1}{6}$	$\frac{1}{6}$	$\frac{1}{6}$	$\frac{1}{6}$	$\frac{1}{6}$
$\frac{1}{4}$		$\frac{1}{4}$		$\frac{1}{4}$	$\frac{1}{4}$
$\frac{1}{3}$		$\frac{1}{3}$		$\frac{1}{3}$	
$\frac{1}{2}$			$\frac{1}{2}$		

$\frac{1}{2} = \frac{2}{4}$ and $\frac{1}{2} = \frac{3}{6}$

Reena can use the remaining cards to complete the puzzle with:

$\frac{1}{2} = \frac{3}{6}$

b) Danny could complete the puzzle in more than one way.

This model helps us to work out sets of equivalent fractions.

$$\overset{\times 3}{\underset{\times 3}{\frac{1}{2} = \frac{3}{6}}} \qquad \overset{\times 2}{\underset{\times 2}{\frac{2}{3} = \frac{4}{6}}} \qquad \overset{\times 2}{\underset{\times 2}{\frac{1}{3} = \frac{2}{6}}}$$

If I write the pairs of equivalent fractions I have found, I think I can see a pattern.

I can find equivalent fractions a different way.

$$\overset{\div 2}{\underset{\div 2}{\frac{4}{6} = \frac{2}{3}}} \quad \text{and} \quad \overset{\div 3}{\underset{\div 3}{\frac{3}{6} = \frac{1}{2}}}$$

Think together

1 **a)** Use the numbers 1, 2, 5 and 10 to make a pair of equivalent fractions.

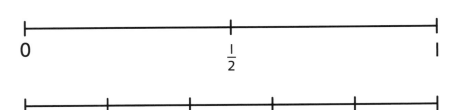

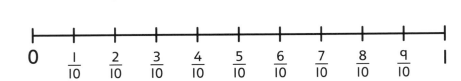

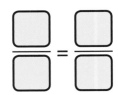

b) Make another pair of equivalent fractions.

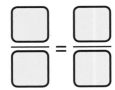

2 Write the missing numerators and denominators for these equivalent fractions.

a) $\dfrac{2}{5} = \dfrac{\boxed{}}{10}$

c) $\dfrac{8}{10} = \dfrac{\boxed{}}{5}$

b) $\dfrac{3}{10} = \dfrac{6}{\boxed{}}$

d) $\dfrac{6}{8} = \dfrac{\boxed{}}{4}$

I used a fraction wall to help me.

18

3 Max and Jamie are trying to find the missing number.

$$\frac{6}{10} = \frac{\square}{5}$$

Max

I multiplied 5 by 2 to get 10, so I need to find what number I multiply 2 by to make 6.

That number is 3.

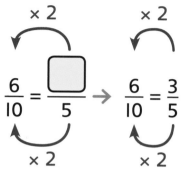

$$\times 2 \qquad \times 2$$
$$\frac{6}{10} = \frac{\square}{5} \rightarrow \frac{6}{10} = \frac{3}{5}$$
$$\times 2 \qquad \times 2$$

Jamie

I divided 10 by 2 to get 5, so I need to divide the numerator 6 by 2.

This gives me 3.

$$\div 2 \qquad \div 2$$
$$\frac{6}{10} = \frac{\square}{5} \rightarrow \frac{6}{10} = \frac{3}{5}$$
$$\div 2 \qquad \div 2$$

Do both Max and Jamie's methods work?

Which method do you prefer?

Use Max or Jamie's method to find the missing numbers.

a) $\dfrac{1}{5} = \dfrac{4}{\square}$

b) $\dfrac{8}{20} = \dfrac{\square}{10}$

c) $\dfrac{\square}{16} = \dfrac{1}{2}$

d) $\dfrac{6}{\square} = \dfrac{2}{3}$

19

→ Practice book 3C p12

Comparing fractions

Discover

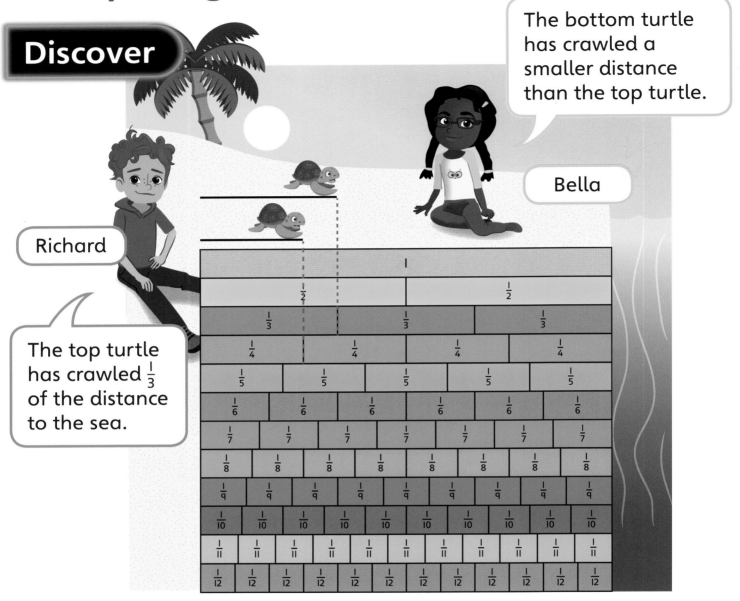

The bottom turtle has crawled a smaller distance than the top turtle.

Bella

Richard

The top turtle has crawled $\frac{1}{3}$ of the distance to the sea.

1 **a)** Write an **inequality statement** to compare how far each turtle has crawled. Use the <, > or = sign. Is Bella correct?

b) After an hour, Richard writes another inequality statement to describe how far the turtles have crawled.

$$\frac{2}{3} < \frac{\square}{4}$$

What could the missing number be?

20

Share

a) The top turtle has crawled $\frac{1}{3}$ of the distance to the sea.

The bottom turtle has crawled $\frac{1}{4}$ of the distance to the sea.

> I think $\frac{1}{4}$ is greater than $\frac{1}{3}$ because 4 is greater than 3.

> I do not think that is correct. I will use the fraction wall to help me compare.

Use the fraction wall to compare thirds and quarters.

$\frac{1}{4}$ is less than $\frac{1}{3}$

$\frac{1}{4} < \frac{1}{3}$

Bella is correct. The bottom turtle has not crawled as far as the top turtle.

b) The fraction has the denominator 4. Use the fraction wall to compare thirds and quarters.

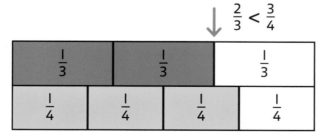

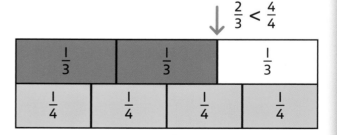

The missing numerator could be 3 because $\frac{2}{3} < \frac{3}{4}$.

The missing numerator could also be 4 because $\frac{2}{3} < \frac{4}{4}$.

Think together

1 **a)** Is $\frac{1}{5}$ greater than or less than $\frac{1}{6}$?

$\frac{1}{5}$ ◯ $\frac{1}{6}$

| $\frac{1}{5}$ | | | | |

| $\frac{1}{6}$ | | | | | |

b) Is $\frac{1}{2}$ greater than or less than $\frac{3}{4}$?

$\frac{1}{2}$ ◯ $\frac{3}{4}$

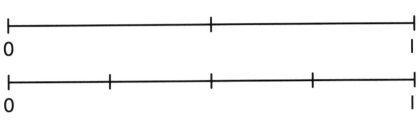

2 Use >, < or = to make these statements true.

a) $\frac{1}{3}$ ◯ $\frac{1}{9}$

b) $\frac{1}{8}$ ◯ $\frac{2}{9}$

c) $\frac{3}{4}$ ◯ $\frac{7}{12}$

1											
$\frac{1}{2}$						$\frac{1}{2}$					
$\frac{1}{3}$				$\frac{1}{3}$				$\frac{1}{3}$			
$\frac{1}{4}$			$\frac{1}{4}$			$\frac{1}{4}$			$\frac{1}{4}$		
$\frac{1}{5}$		$\frac{1}{5}$		$\frac{1}{5}$		$\frac{1}{5}$			$\frac{1}{5}$		
$\frac{1}{6}$		$\frac{1}{6}$		$\frac{1}{6}$		$\frac{1}{6}$		$\frac{1}{6}$		$\frac{1}{6}$	
$\frac{1}{7}$	$\frac{1}{7}$		$\frac{1}{7}$		$\frac{1}{7}$		$\frac{1}{7}$		$\frac{1}{7}$		$\frac{1}{7}$
$\frac{1}{8}$	$\frac{1}{8}$	$\frac{1}{8}$		$\frac{1}{8}$	$\frac{1}{8}$		$\frac{1}{8}$	$\frac{1}{8}$		$\frac{1}{8}$	
$\frac{1}{9}$	$\frac{1}{9}$	$\frac{1}{9}$	$\frac{1}{9}$	$\frac{1}{9}$		$\frac{1}{9}$	$\frac{1}{9}$	$\frac{1}{9}$		$\frac{1}{9}$	
$\frac{1}{10}$	$\frac{1}{10}$	$\frac{1}{10}$	$\frac{1}{10}$	$\frac{1}{10}$	$\frac{1}{10}$	$\frac{1}{10}$	$\frac{1}{10}$	$\frac{1}{10}$	$\frac{1}{10}$		
$\frac{1}{11}$	$\frac{1}{11}$	$\frac{1}{11}$	$\frac{1}{11}$	$\frac{1}{11}$	$\frac{1}{11}$	$\frac{1}{11}$	$\frac{1}{11}$	$\frac{1}{11}$	$\frac{1}{11}$	$\frac{1}{11}$	
$\frac{1}{12}$	$\frac{1}{12}$	$\frac{1}{12}$	$\frac{1}{12}$	$\frac{1}{12}$	$\frac{1}{12}$	$\frac{1}{12}$	$\frac{1}{12}$	$\frac{1}{12}$	$\frac{1}{12}$	$\frac{1}{12}$	$\frac{1}{12}$

I will use the fraction wall to help me work out the answers.

3

$\frac{1}{8}$		$\frac{1}{8}$		$\frac{1}{8}$		$\frac{1}{8}$		$\frac{1}{8}$		$\frac{1}{8}$		$\frac{1}{8}$		$\frac{1}{8}$	
$\frac{1}{10}$		$\frac{1}{10}$	$\frac{1}{10}$	$\frac{1}{10}$		$\frac{1}{10}$	$\frac{1}{10}$		$\frac{1}{10}$	$\frac{1}{10}$		$\frac{1}{10}$	$\frac{1}{10}$	$\frac{1}{10}$	
$\frac{1}{12}$	$\frac{1}{12}$	$\frac{1}{12}$	$\frac{1}{12}$	$\frac{1}{12}$	$\frac{1}{12}$	$\frac{1}{12}$	$\frac{1}{12}$	$\frac{1}{12}$	$\frac{1}{12}$	$\frac{1}{12}$	$\frac{1}{12}$				

a) Use the fraction wall to help you complete the statement.

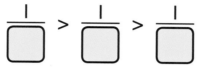

$$\frac{1}{\Box} > \frac{1}{\Box} > \frac{1}{\Box}$$

b) How do the bars help you to explain your reasoning?

4 a) Isla says that $\frac{7}{10}$ is greater than $\frac{4}{5}$, because 7 is greater than 4, and 10 is greater than 5.
Is Isla correct? Explain your answer.

CHALLENGE

b) What numbers could go in the missing boxes?

$$\frac{4}{5} > \frac{\Box}{6}$$

$$\frac{4}{5} > \frac{6}{\Box}$$

Do not forget that you can use the fraction wall!

23

Comparing and ordering fractions

Discover

Drag the number cards to complete the puzzles. You can only use each card once!

Complete the puzzle to move to the next level.

$$\frac{1}{8} < \frac{\square}{\square} < \frac{\square}{\square}$$

Cards: 4, 5, 2, 10

Complete the puzzle to move to the next level.

$$\frac{\square}{4} < \frac{3}{4} \qquad \frac{4}{8} > \frac{4}{\square}$$

$$\frac{\square}{10} = \frac{\square}{8}$$

Cards: 4, 5, 2, 10

Olivia

Luis

1 **a)** Complete the puzzle for Luis.

b) Complete the puzzle for Olivia.

Share

I will use number lines to help me work it out!

a) Start by completing the equivalent fraction.
There is only one way to do this using Luis's cards.

$\frac{5}{10} = \frac{4}{8}$

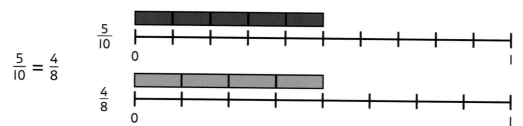

Now use the other cards to complete the rest of the puzzles.

$\frac{2}{4} < \frac{3}{4}$

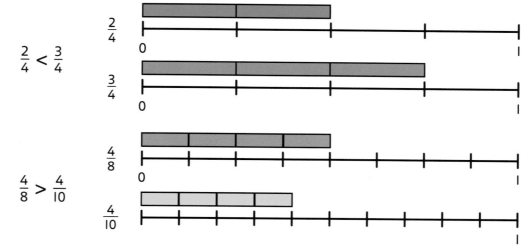

$\frac{4}{8} > \frac{4}{10}$

b)

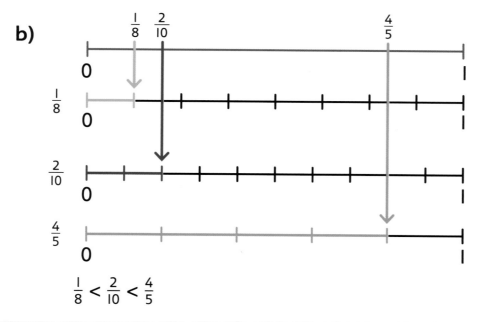

First I will make a list of the fractions I can make with Olivia's cards. Then I will order the fractions using a number line.

$\frac{1}{8} < \frac{2}{10} < \frac{4}{5}$

25

Think together

1 What numbers could go in the boxes?

a) $\dfrac{3}{5} > \dfrac{\Box}{5}$

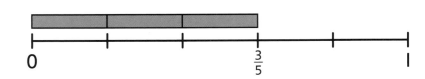

b) $\dfrac{5}{6} < \dfrac{5}{\Box}$

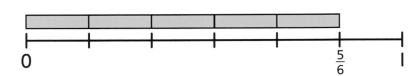

I wonder if $\frac{5}{6}$ is less than one whole.

2 Order these fractions from smallest to largest.

Use the number lines to help.

a) $\dfrac{8}{8}$ $\dfrac{2}{8}$ $\dfrac{6}{8}$

b) $\dfrac{1}{4}$ $\dfrac{1}{2}$ $\dfrac{1}{3}$

c) $\dfrac{1}{2}$ $\dfrac{3}{4}$ $\dfrac{1}{3}$

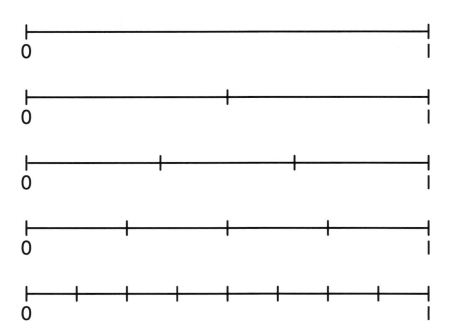

3 **a)** What numbers should go in the boxes?

CHALLENGE

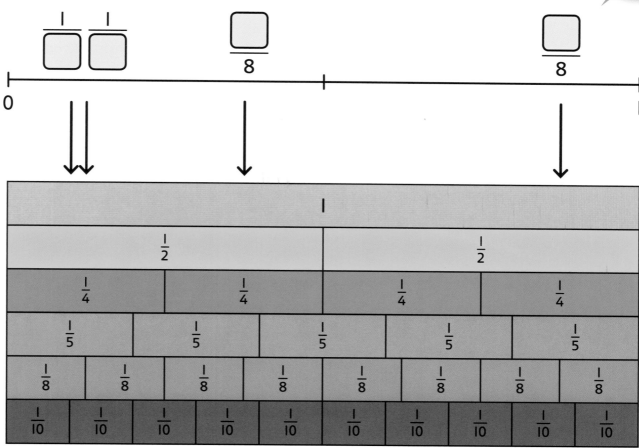

b) Write the four fractions in order, starting with the greatest.

How will you decide what the missing numerators and denominators should be?

27

Adding fractions

Discover

Zac

Isla

1 **a)** Altogether, what fraction of pizza is left in the boxes?

b) Use a number line to show your answer.

Share

a) The first box has 4 tenths or $\frac{4}{10}$ of a pizza left.

The second box has 3 tenths or $\frac{3}{10}$ of a pizza left.

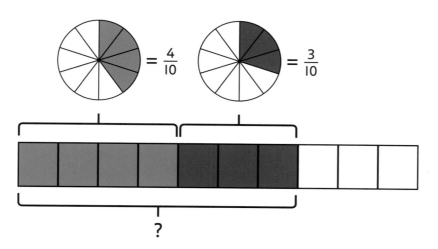

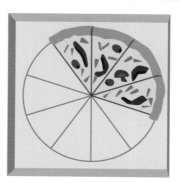

4 tenths + 3 tenths = 7 tenths

$\frac{4}{10} + \frac{3}{10} = \frac{7}{10}$ So, altogether $\frac{7}{10}$ of a pizza is left in the boxes.

b)

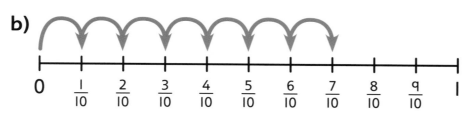

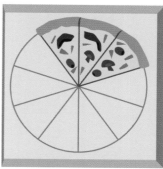

I jumped $\frac{1}{10}$ at a time.

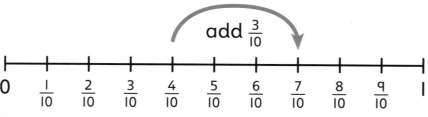

add $\frac{3}{10}$

I started at $\frac{4}{10}$ and jumped $\frac{3}{10}$ in one go.

I wonder if you get the same result if you start with $\frac{3}{10}$ and add $\frac{4}{10}$.

Think together

1 Add these fractions.

a) $\frac{4}{8} + \frac{1}{8} = \boxed{} \atop \boxed{}$

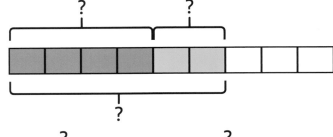

b) $\frac{4}{9} + \frac{2}{9} = \boxed{} \atop \boxed{}$

c) $\frac{2}{6} + \frac{4}{6} = \boxed{} \atop \boxed{}$

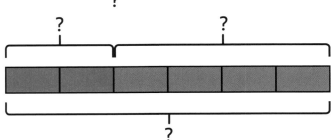

2 Add these fractions.

a) $\frac{3}{5} + \frac{1}{5} = \boxed{} \atop \boxed{}$

b) $\frac{3}{5} + \frac{2}{5} = \boxed{} \atop \boxed{}$

c) $\frac{1}{6} + \frac{3}{6} = \boxed{} \atop \boxed{}$

d) $\frac{5}{12} + \frac{1}{12} = \boxed{} \atop \boxed{}$

I will use a fraction strip to check my answers.

30

3 **a)** What calculation is shown by this fraction strip and number line?

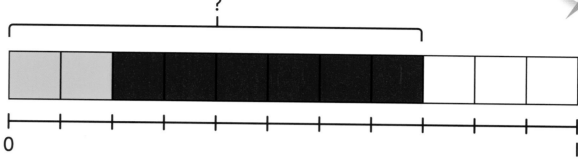

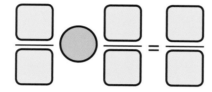

b) The answer is 'a fraction less than I'. What is the question?

Find three possible answers.

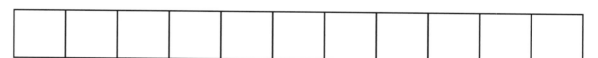

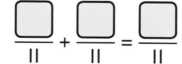 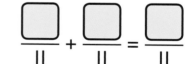

I think there must be some more answers I could find, too.

I wonder if I can add three fractions together and still make an answer that is less than I.

31

Subtracting fractions

Discover

1 a) How much fuel will be left after the journey home?

b) If the fuel was full before the journey home, what would the answer be?

Share

a) The fuel tank is $\frac{5}{8}$ full. The journey home will use $\frac{3}{8}$ of the fuel.

5 eighths − 3 eighths = 2 eighths

$$\frac{5}{8} - \frac{3}{8} = \frac{2}{8}$$

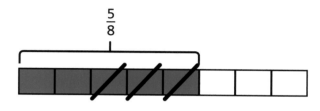

I drew a bar and shaded in $\frac{5}{8}$. Then I crossed out $\frac{3}{8}$.

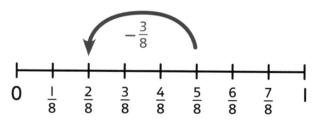

I jumped back $\frac{3}{8}$ on a number line.

There will be $\frac{2}{8}$ of the fuel left after the journey home.

b) You start with $\frac{8}{8}$ and use $\frac{3}{8}$.

$$1 - \frac{3}{8} = \frac{8}{8} - \frac{3}{8} = \frac{5}{8}$$

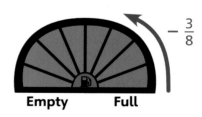

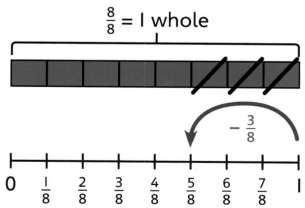

If the fuel was full before the journey home, there would be $\frac{5}{8}$ of the fuel left after the journey home.

Think together

1 Complete these subtractions.

a) $\dfrac{7}{8} - \dfrac{5}{8} = \dfrac{\square}{\square}$

b) $1 - \dfrac{3}{10} = \dfrac{\square}{\square}$

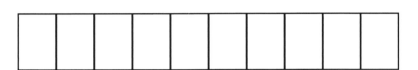

I can remember how to write a whole as a fraction.

2 Work out these subtractions.

a) $\dfrac{6}{11} - \dfrac{2}{11} = \dfrac{\square}{\square}$

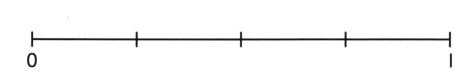

b) $1 - \dfrac{1}{4} = \dfrac{\square}{\square}$

0   1

3 **a)** Complete the calculation.

CHALLENGE

$$\frac{\boxed{}}{7} - \frac{\boxed{}}{7} = \frac{2}{7}$$

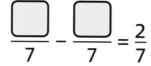

> I think I can find more than one answer.

b) The difference between two fractions is $\frac{3}{10}$.

What could the fractions be?
Use the number line to help you.

> What does 'the difference' mean?

0    $\frac{1}{10}$ $\frac{2}{10}$ $\frac{3}{10}$ $\frac{4}{10}$ $\frac{5}{10}$ $\frac{6}{10}$ $\frac{7}{10}$ $\frac{8}{10}$ $\frac{9}{10}$ 1

35

Problem solving – adding and subtracting fractions

Discover

We used $\frac{1}{10}$ of the food on Monday and $\frac{3}{10}$ of the food on Tuesday.

1 **a)** What fraction of the food is left in the box?

b) Has more food been used or left in the box? How much more?

Share

a)

I will start by working out the food used on Monday and Tuesday.

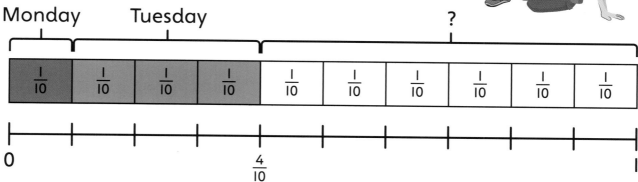

$\frac{1}{10} + \frac{3}{10} = \frac{4}{10}$, so $\frac{4}{10}$ of the food has been used.

$1 - \frac{4}{10} = \frac{6}{10}$, so $\frac{6}{10}$ of the food is left in the box.

b)

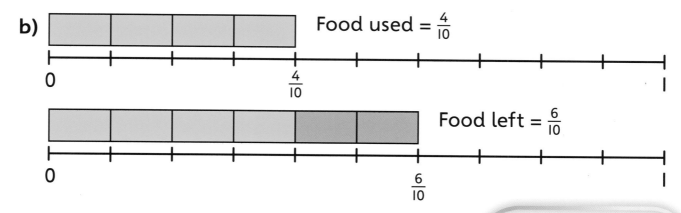

Food used = $\frac{4}{10}$

Food left = $\frac{6}{10}$

More of the food is left than has been used.

$\frac{6}{10} - \frac{4}{10} = \frac{2}{10}$

$\frac{2}{10}$ more of the food is left in the box than has been used.

I used a number line to work out the difference.

Think together

1 Sofia and Amal go on a journey. They walk $\frac{5}{8}$ of the journey and ski $\frac{1}{8}$ of the journey. Then they stop for a rest.

What fraction of the journey is left?

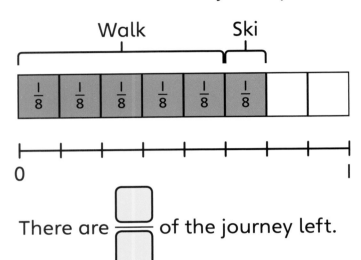

Show your workings clearly. It is important to show every step even if you can see the answer straight away!

There are $\frac{\square}{\square}$ of the journey left.

2 $\frac{3}{9}$ of the tents at the polar camp are blue, $\frac{1}{9}$ of the tents are red. The rest of the tents are yellow.

a) What fraction of the tents are yellow?

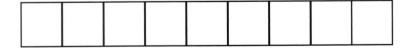

$\frac{\square}{\square}$ of the tents are yellow.

b) Are there more yellow or blue tents? Explain your answer.

There are $\frac{\square}{\square}$ more _____ tents than _____ tents.

3 Alex ate $\frac{1}{5}$ of a packet of raisins. Max ate $\frac{1}{5}$ of a packet more raisins than Alex.

CHALLENGE

a) What fraction of the packet did Alex and Max eat altogether?

First, I will work out what fraction of the packet Max ate.

I will use a fraction strip to represent the fraction that both the children ate.

Max and Alex ate $\frac{}{}$ of the packet of raisins altogether.

b) What fraction of the packet was left?

$\frac{}{}$ of the packet of raisins was left.

39

→ Practice book 3C p27

Problem solving – fractions of measures

Discover

1 **a)** Amelia and Lee spend the rest of the £20 note on sandwiches. What fraction is this?

b) How much money will they have left to spend on sandwiches?

Share

a) The fraction they will spend on juice and fruit in total is $\frac{2}{10} + \frac{4}{10} = \frac{6}{10}$.

> I will use a fraction strip to work out the total fraction spent on juice and fruit, and then see what is left.

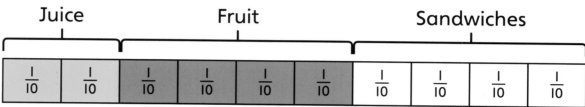

The fraction left to spend on sandwiches is

$1 - \frac{6}{10} = \frac{4}{10}$.

Lee and Amelia spend $\frac{4}{10}$ of £20 on sandwiches.

> First I will find $\frac{1}{10}$ of £20, then I will use a fraction strip to represent the problem.

b) $\frac{1}{10}$ of £20 = £20 ÷ 10 = £2

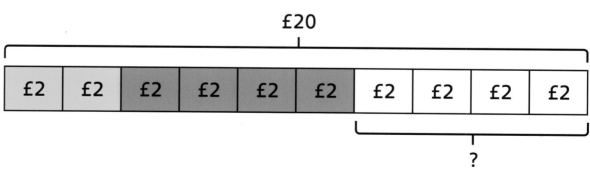

$\frac{4}{10}$ of £20 = 4 × £2 = £8

They will have £8 left to spend on sandwiches.

> I wonder how I could check the answer is correct.

Think together

1 Miss Hall buys 12 kg of oranges for her class.

They use $\frac{4}{6}$ of the oranges to make orange juice.

The children eat $\frac{1}{6}$ of the oranges.

12 kg

a) What fraction of the oranges are left?

b) How many kilograms of oranges are left?

2 Olivia separates items of food onto two tables.

She puts the larger amount of each item on the large table.
She puts the smaller amount of each item on the small table.

Which items should go on each table?

A $\frac{5}{9}$ of 1 kg of chocolates

B $\frac{4}{9}$ of 1 kg of chocolates

C $\frac{3}{4}$ of 3 kg of strawberries

D $\frac{3}{5}$ of 3 kg of strawberries

E $\frac{7}{12}$ of a bag of nuts

F $\frac{7}{10}$ of a bag of nuts

I wonder if I can work some of these out without doing any calculations.

3 Aki and his mum go for a long walk. They walk 16 km in total. They walk $\frac{1}{8}$ of the distance, have lunch and then walk $\frac{5}{8}$ of the distance.

a) How far do they have left to walk?

b) Find two ways to answer this question. Which method do you prefer?

24 km

I am going to find how many kilometres they have walked so far and then do a subtraction.

I will work out the fraction they have left to walk first.

43

→ Practice book 3C p30

End of unit check

1 Which of these fractions is equivalent to $\frac{2}{3}$?

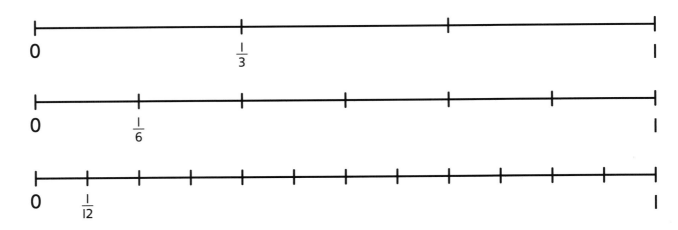

A $\frac{8}{12}$ **B** $\frac{9}{12}$ **C** $\frac{5}{6}$ **D** $\frac{4}{12}$

2 Look at these equivalent fractions. Which number is not used to complete them?

$\frac{1}{4} = \frac{\Box}{8}$ $\frac{6}{20} = \frac{\Box}{10}$ $\frac{4}{10} = \frac{2}{\Box}$

A 2   **B** 3   **C** 4   **D** 5

3 Complete the sentence $\frac{1}{4} > \frac{1}{\Box}$

A 2   **B** 3   **C** 4   **D** 5

4 Which fraction cannot go into any of these fraction sentences?

$$\frac{3}{5} + \frac{\square}{\square} = 1 \qquad \frac{7}{10} + \frac{\square}{\square} = 1 \qquad \frac{\square}{\square} + \frac{1}{8} = 1$$

A $\frac{3}{5}$ **B** $\frac{7}{8}$ **C** $\frac{2}{5}$ **D** $\frac{3}{10}$

5 Jake painted $\frac{1}{6}$ of the wall on Monday and $\frac{3}{6}$ of the wall on Tuesday. What fraction of the wall was not painted?

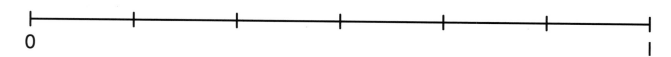

0   1

A $\frac{1}{3}$ **B** $\frac{4}{6}$ **C** $\frac{2}{3}$ **D** $\frac{1}{2}$

6 Which numbers could be the missing numerator?

$\frac{5}{8}$ of 40 kg $< \dfrac{\square}{5}$ of 40 kg

A 3   **C** 2

B 4   **D** 5

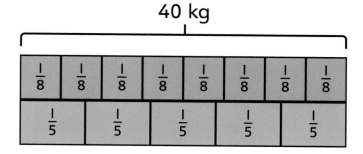

40 kg

| $\frac{1}{8}$ | $\frac{1}{8}$ | $\frac{1}{8}$ | $\frac{1}{8}$ | $\frac{1}{8}$ | $\frac{1}{8}$ | $\frac{1}{8}$ | $\frac{1}{8}$ |

| $\frac{1}{5}$ | $\frac{1}{5}$ | $\frac{1}{5}$ | $\frac{1}{5}$ | $\frac{1}{5}$ |

7 Write these fractions in order, starting with the smallest.

$\frac{1}{2}$ $\frac{1}{9}$ $\frac{2}{3}$ $\frac{3}{4}$ $\frac{1}{12}$

45

→ Practice book 3C p33

Unit 11
Time

In this unit we will ...

⚡ Learn about hours, days, months and years

⚡ Estimate times

⚡ Tell the time to the nearest minute

⚡ Calculate start and end times

⚡ Solve time problems

Do you remember how to count the number of minutes past or to an o'clock time?

5 minutes
10 minutes
15 minutes
20 minutes
25 minutes

5 minutes
10 minutes
15 minutes
20 minutes
25 minutes

We will be using some maths words. Do you recognise any of these?

month   year   midnight   midday

am   pm   duration   estimate

consecutive   hour   minute   second

past   to   start   end

duration   digital clock   analogue clock

How do you know what the time is?

Months and years

Discover

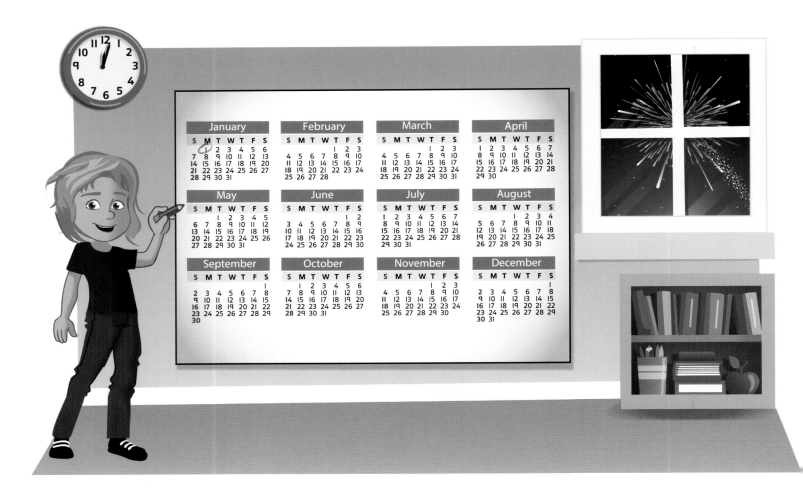

1 **a)** How many months are in a year?

How many days are in each month?

b) How many days are in a year?

Share

a) There are 12 months in a year.

Look at the last day of each month. It shows how many days are in that month.

> A year is the time it takes for planet Earth to travel once around the sun.

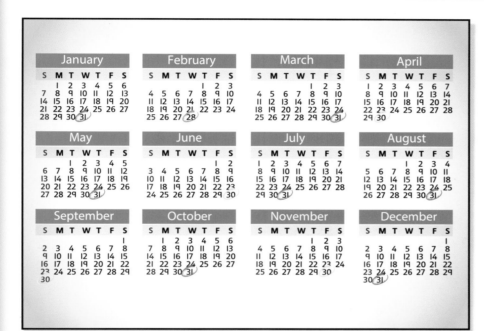

January, March, May, July, August, October and December have 31 days.

April, June, September and November have 30 days.

February has 28 days (29 days in a leap year).

> I can use my knuckles to help me remember! The months on my knuckles have 31 days.

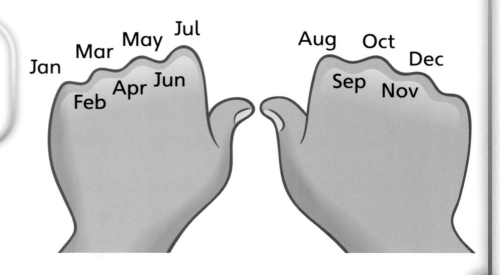

b) The number of days in each month helps us to find out how many days are in a whole year.

One month has **28** days.

Four months have 30 days. 4 × 30 = 120 days

Seven months have 31 days. 7 × 31 = 217 days

28 + 120 + 217 = 365 days. There are 365 days in a year.

Think together

1

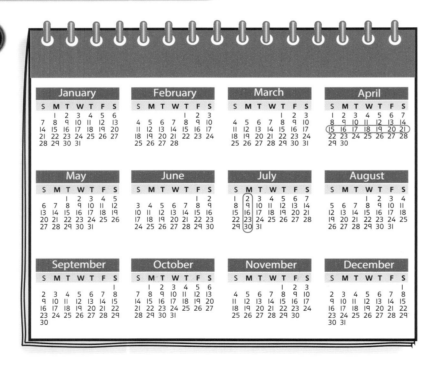

Some years are leap years. They have an extra day.

a) Use the calendar to help you count forwards one week from 14 April. Count in days. 15 April is day 1.

b) Use the calendar to help you count backwards four weeks from 30 July. Count in weeks.

2 It is the 188th day of the year.

It is not a leap year.

How many days are left in the year?

365 − 188 = ☐

There are ☐ days left in the year.

one year

365 days	
188 days	? days

A leap year has 366 days.

CHALLENGE

3 It takes nearly $365 \frac{1}{4}$ days for the Earth to travel once around the Sun. This means the calendar needs to be adjusted almost every four years.

How do you think the calendar is adjusted for the extra quarter days?

We cannot have a quarter of a day at the end of a year. I think this has something to do with leap years!

365 days	$\frac{1}{4}$
365 days	$\frac{1}{4}$
365 days	$\frac{1}{4}$
365 days	$\frac{1}{4}$

51

Hours in a day

Discover

1. a) When does a day start and end?

   b) How many hours are there in one day?

Share

a)

midnight   midday   midnight

12 1 2 3 4 5 6 7 8 9 10 11 12 1 2 3 4 5 6 7 8 9 10 11 12

The start of the day is 12 o'clock at night. This is called midnight.

The middle of the day is 12 o'clock midday, or noon.

The end of the day is the next midnight. This is when a new day begins.

> Can it really be the start of a new day at midnight, even though it is dark?

> We start counting the 24 hours in a day at midnight.
>
> They last until the next midnight when we start another new day.

b) There are 12 hours from midnight until midday.

There are another 12 hours from midday until the next midnight.

12 + 12 = 24

There are 24 hours in one day.

Think together

1   Use a clock face to show how long a day lasts.

I am going to start with both hands showing midnight!

One day lasts for ☐ hours.

It starts at _____ and ends at _____ the following day.

2

Think carefully before you answer these!

a)   How many times does the minute hand travel all the way around the clock in one day?

The minute hand travels all the way around ☐ times.

b)   How many times does the hour hand travel all the way around the clock in one day?

The hour hand travels all the way around ☐ times.

3 It is 1 o'clock in the morning on Tuesday.

a) Without counting each separate hour, how many hours are there until 3 o'clock in the morning on Wednesday?

Tuesday morning

Wednesday morning

b) How many hours are there from 3 o'clock Wednesday morning until 3 o'clock in the morning on the Wednesday one week later?

I need to know the number of days in a week to work this out!

I am going to use multiplication to help me.

55

→ Practice book 3C p38

Estimating time

Discover

1 **a)** How can we estimate the time, even though the minute hand has fallen off?

b) What time do you estimate it is?

Share

a) The hour hand takes one hour to move between two numbers.

We can use its position to estimate what the time is.

At an o'clock time, the hour hand points directly at a number.

At a half-past time, it has moved to half-way between two numbers.

The hour hand has now passed the half-way point. I will use this to help me estimate!

b) The hour hand is pointing half-way between the 10 and the 11.

We can estimate that the time is about half past 10.

Think together

1 Use the position of the hour hand to estimate the time.

It is between the 6 and the 7. The time is after 6 o'clock and before 7 o'clock.

It has not reached the half-way point. It is before half past 6.

The time is about _____ .

2 Here is part of a clock face.

Point to where you think the hour hand should be if the time is quarter to 4.

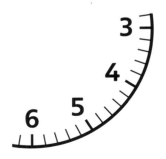

I am going to use my knowledge of fractions to help!

3 **a)** There are ☐ minutes in one hour.

Explain how you know this.

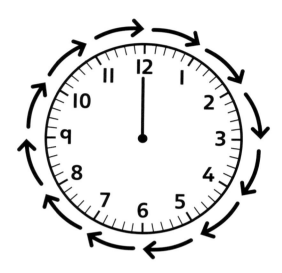

b) We can use the small marks on the clock face to help us estimate the time.

The hour hand moves five small marks every hour.

How many minutes is each small mark worth?

I need to know the number of minutes in an hour to work this out.

59

Telling time to 5 minutes

Discover

① **a)** How long has it been since the last train left?

b) How long will it be until the next train leaves?

Share

a) All trains leave on the hour. This means every train leaves at an o'clock time.

Look at the minute hand. It has moved past 11 numbers since the last o'clock time.

There are 5 minutes between each number on a clock.

11 × 5 = 55

It has been 55 minutes since the last train left.

b)

5 minutes to   55 minutes to

10 minutes to   50 minutes to

15 minutes to   45 minutes to

20 minutes to   40 minutes to

25 minutes to   35 minutes to

30 minutes to

I count backwards from an o'clock time to find the minutes to the next hour.

The next train leaves at 12 o'clock.

The time is 5 minutes to 12, so it will be 5 minutes until the next train leaves.

Think together

1 The clock at a bus station looks like this. What time is it?

This clock has Roman numerals as well as normal numerals.

 4 o'clock

five minutes to 4

?   ?   ?

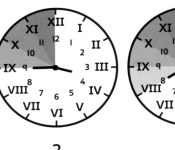

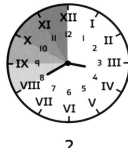

There are ☐ minutes until ☐ o'clock.

The time is _____ .

I will count backwards from 4 o'clock.

2 Look at the clocks below. What time does each one show?

I think counting in 5s is going to be useful here.

3 Use the clock face to help you answer these questions.

CHALLENGE

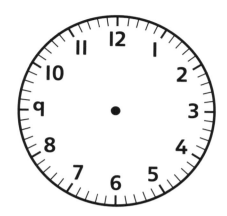

a) Where might the minute and hour hands be if the time is 'something' past 6?

b) Where might the minute and hour hands be at 'something' to 10?

c) Where might the hands be if the time is twenty to 'something'?

I can say which half of the clock the minute hand is in.

I can say which two numbers the hour hand is pointing between!

→ Practice book 3C p44

Telling time to the minute ❶

Discover

❶ **a)** How many minutes past 10 was this photo taken?

b) Is there another way to say this time?

Share

a) There are 60 minutes in 1 hour.

> I will count in jumps of 5s, and then 1s, to work out how many minutes past 10 o'clock it is.

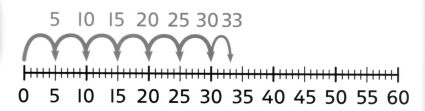

5 minutes
10 minutes
15 minutes
20 minutes
25 minutes
30 minutes

> I can show these jumps on a number line!

The photo was taken at 33 minutes past 10.

b) We can also count backwards from 11 o'clock.

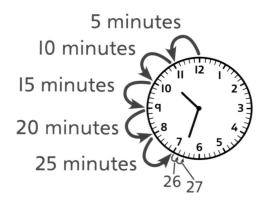

5 minutes
10 minutes
15 minutes
20 minutes
25 minutes

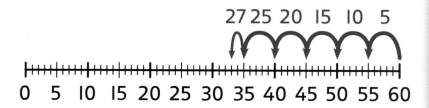

> I can show this as a subtraction.

Minutes in 1 hour – minutes past = minutes to

60 – 33 minutes past = 27 minutes to

Another way to say this time is 27 minutes to 11.

Think together

1 What time does each clock show?

I will count in 5s and Is.

a)

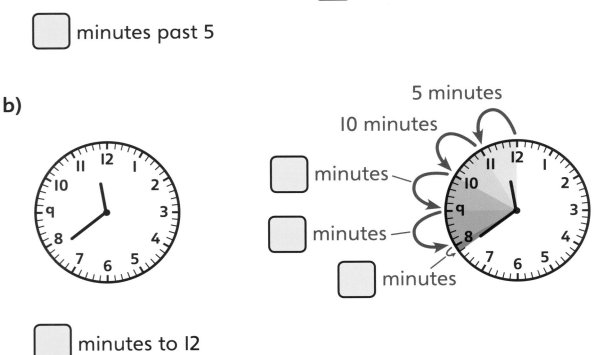

5 minutes

10 minutes

☐ minutes

☐ minutes

☐ minutes

☐ minutes

☐ minutes past 5

b)

5 minutes

10 minutes

☐ minutes

☐ minutes

☐ minutes

☐ minutes to 12

2 What time is it?

a)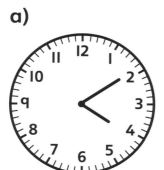

b)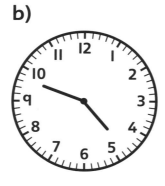

c)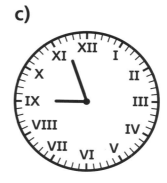

d)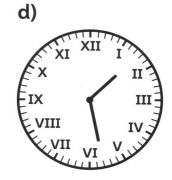

a) [ ] minutes past 4

b) [ ] minutes to 5

c) [ ] minutes to [ ]

d) [ ] minutes past [ ]

3 This clock has no numbers on it.

Read the time to the nearest minute. Explain how you did it.

CHALLENGE

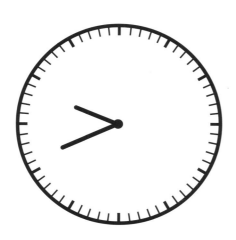

I can think of two ways to say this time.

67

→ Practice book 3C p47

Telling time to the minute 2

Discover

1 **a)** What time should the clock on the wall show?

b) Is it the morning or the evening? How do you know?

Share

a)

I wonder what the numbers on the **digital** clock represent.

The first number shows the hour. The second number shows the number of minutes past that hour. This is always two digits.

The digital clock shows 54 minutes past 8.

We say this as '6 minutes to 9'.

The clock on the wall should show the time like this.

b) The letters 'am' and 'pm' show what time of day it is.

These letters come from Latin.

Ante meridiem (am) means before midday (from midnight until midday).

Post meridiem (pm) means after midday (from midday until midnight).

It is morning because the digital clock says 'am'.

Think together

1. Which clocks show the same times as each other?

A

C

B

D

1

3

2

4

Clock A shows the same time as clock ☐.

Clock B shows the same time as clock ☐.

Clock C shows the same time as clock ☐.

Clock D shows the same time as clock ☐.

2 It is the evening.

Which digital clock shows the same time as the clock face?

AM

A

AM

C

PM

B

PM

D

The clock face shows ☐ minutes past ☐ .

Evening is shown by the letters _____ .

Digital clock _____ shows the same time as the clock face.

3 How many minutes is it **to** the hour?

CHALLENGE

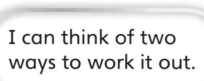

PM

I think this shows minutes past the hour.

I can think of two ways to work it out.

→ Practice book 3C p50

Telling time to the minute ③

Discover

1 **a)** Were the woodpecker and the fox seen in the morning or the afternoon?

   b) The badger was spotted in the evening. What time should the digital clock show?

Share

a) The 24-hour clock splits the day into 24 hours, from 00:00 (12:00 am) to 23:59 (11:59 pm).

This shows if a time is before or after midday, without using am or pm.

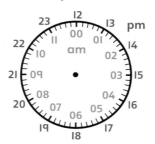

For example:

6:25 am is written as 06:25

4:45 pm is written as 16:45

> I have noticed that all 24-hour clock times have four digits, so some numbers have zeros in front of them.

> I have spotted something too!
>
> I can work out the 24-hour clock times by adding 12 hours to the times from 1 pm onwards.

07:32 is the same as 7:32 am.

The woodpecker was seen in the morning.

13:24 is the same as 1:24 pm.

The fox was seen in the afternoon.

b) The badger was spotted at half past 10 in the evening.

The digital clock shows this time as a 24-hour clock time. It should show 22:30.

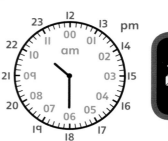

73

Think together

1 Mo spots a robin at 4:46 pm.

How should he write this as a 24-hour clock time?

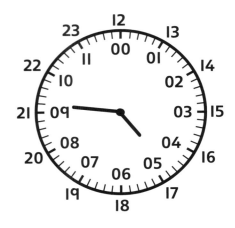

I am going to try adding 12 to change 4:46 pm into a 24-hour clock time.

4 + 12 = ☐

The robin was spotted at ☐ : ☐ .

2 Write these times as 24-hour clock times.

Type of animal	Time it was seen (12-hour)	Time it was seen (24-hour)
Blackbird	9:16 am	
Bat	9:50 pm	
Owl	11:15 pm	

3 Only one person is correct. Who is it and why?

To change this time into a 24-hour clock time, I need to add 12 to make it 24:15.

To change this time into a 24-hour clock time, I only need to remove the 'am' and write it as 8:25.

My clock shows 2:50 in the afternoon as a 24-hour clock time.

12:15 pm

Richard

8:25 am

Jamilla

14:50 pm

Ambika

I will try to remember which times I add 12 to when making 24-hour times.

→ Practice book 3C p53

Finding the duration

Discover

1 **a)** How long does it take the farmer to plough the field?

b) When he has finished, how many more minutes go by before he has a cup of tea?

Share

a) The time something takes is called its **duration**.

16:23

16:51

I am going to count forwards from the start time until I reach the end time.

23 + 28 = 51

+ 7 minutes   + 21 minutes

16:00   17:00

It takes the farmer 28 minutes to plough the field.

b)

 16:51

 17:00

 17:19

+ 9 minutes   + 19 minutes

I will use a number line to find the answer.

+ 9 minutes   + 19 minutes

16:50   17:00   17:10   17:20

9 + 19 = 28 minutes

Another 28 minutes go by before the farmer stops for a cup of tea.

Think together

1 A farmer starts milking the cows at 05:13.

She finishes at 05:57.

How long does it take?

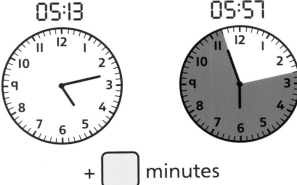

05:13

05:57

+ ☐ minutes

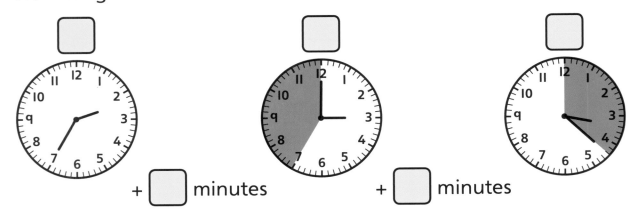

05:00 05:05 05:10 05:15 05:20 05:25 05:30 05:35 05:40 05:45 05:50 05:55 06:00

It takes the farmer ☐ minutes to milk the cows.

2 A lorry driver arrives to collect vegetables at 35 minutes past 2 in the afternoon.

He leaves at 22 minutes past 3.

How long was he at the farm for?

☐   ☐   ☐

+ ☐ minutes   + ☐ minutes

☐ + ☐ = ☐

The lorry driver was at the farm for ☐ minutes.

3

I started collecting the eggs at 16:32. I finished at 17:55.

How long did it take the farmer to collect all the eggs?

Show two ways of finding the answer.

I will try counting up.

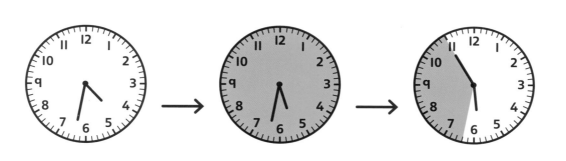

I wonder if there is a quicker way when the answer will be more than one hour.

79

Comparing duration

Discover

1 **a)** Which activity takes the longest?

b) The Arts and Crafts activity takes less time than the Poetry Workshop.

It takes more time than the Meet the Author event.

When might the Arts and Crafts activity finish?

Share

I will count the minutes of each activity using a clock face.

a)

Meet the Author:
9:06 to 10:00 = 54 minutes

+ 54 minutes

Songs and Stories:
11:35 to 12:20 = 45 minutes

+ 25 minutes   + 20 minutes

25 + 20 = 45 minutes

Poetry Workshop:
13:40 to 14:40 = 60 minutes

+ 60 minutes

Remember 60 minutes make 1 hour.

I will show the number of minutes on a number line.

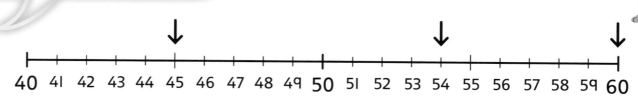

45 minutes < 54 minutes   54 minutes < 60 minutes

The Poetry Workshop activity takes the longest.

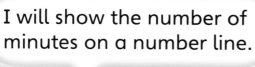

b)

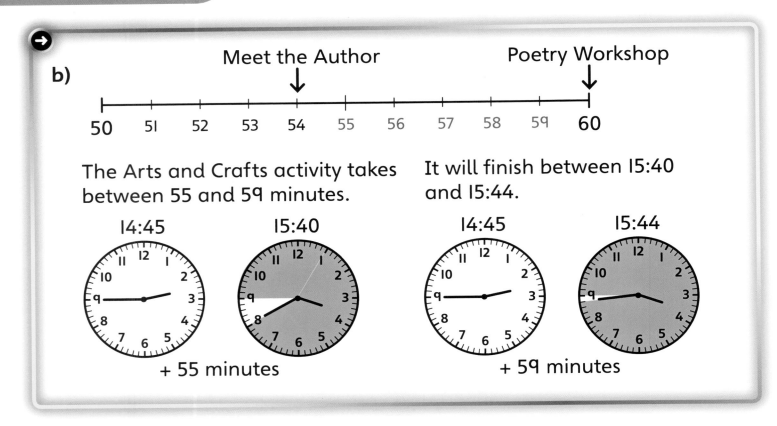

Meet the Author

Poetry Workshop

The Arts and Crafts activity takes between 55 and 59 minutes.

It will finish between 15:40 and 15:44.

14:45   15:40

+ 55 minutes

14:45   15:44

+ 59 minutes

Think together

1   Here are the times for two library events. Which event takes longer?

Library event	Start	End
Story Time	16:40	17:09
Make a Book	16:45	17:11

The second event starts 5 minutes later. I can use this to help me see which takes longer.

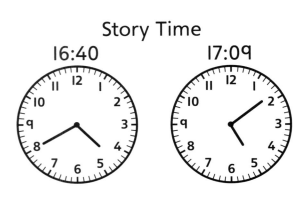

Story Time

16:40   17:09

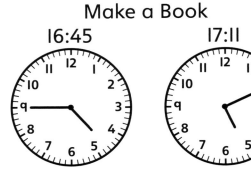

Make a Book

16:45   17:11

_____ takes longer.

2 Which of these durations is the longest?

a)

from 8:12 am
until 8:48 am

b)

from 8:43 am
until 8:57 am

c)

from 8:28 am
until 9:03 am

3 Order these durations, from shortest time to longest time.

A from 4:28 pm until 5:35 pm

B 64 minutes

C I hour and 5 minutes

D from 16:37 until 17:40

I am going to work out
A and D to start with ...

Changing all the times
into minutes might help
me to compare them.

83

→ Practice book 3C p59

Finding start and end times

Discover

I want to go on the big wheel at 5 past 4.

1. a) Max has just started queuing for the dodgems. When will he get on?

   b) When should Olivia start queuing for her ride?

Share

a)

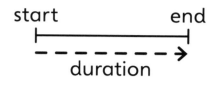

I will use the start time and the duration to help me find the end time!

Start + Duration = End

 + DODGEMS Queuing time 6 minutes =

The time is now 2:47 pm.

47 + 6 = 53

Max will get on the dodgems at 2:53 pm.

b)

I am going to use two subtractions because it crosses an o'clock time.

First, subtract five minutes to get to 4 o'clock. Then, subtract the rest of the duration from 60.

End time   Duration   Start time

End time   −   Duration   =   Start time

 − BIG WHEEL Queuing time 30 minutes =

Olivia should start queuing at 3:35 pm.

85

Think together

1 It is 4:36 pm. If Max and Olivia start queuing now, when can they go on the carousel?

CAROUSEL QUEUING TIME:

14 minutes

Start time   Duration   End time

4:36 pm   14 minutes   _____

Max and Olivia can go on the carousel at _____ .

2 The time now is 10:09 am. Luis has been queuing for 25 minutes.

What time did he start queuing?

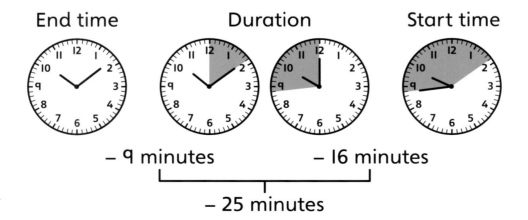

End time   Duration   Start time

– 9 minutes   – 16 minutes

– 25 minutes

Luis started queueing at _____ .

3 **a)** The queue for the helter-skelter takes 12 minutes.

Bella began queuing at 8 minutes to 3.
What time will she go on the ride?

Use the number line to work this out.
Think about the number of minutes past the hour.

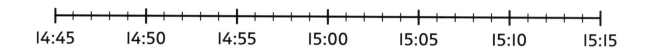

14:45   14:50   14:55   15:00   15:05   15:10   15:15

Bella will go on the helter-skelter at _____ .

b) The journey home takes 59 minutes.

If the Khan family leave at 5.35 pm, what time will they get home?

They will get home at _____ .

Explain your method.

I wonder if there are different ways to find the answers to these questions.

→ Practice book 3C p62

Measuring time in seconds

Discover

1 a) How long have Richard and Amelia been playing?

  b) What else could Lee use to measure time in **seconds**?

Share

a) Seconds are used to measure short periods of time.

60 seconds = 1 minute

Richard has been playing for 50 seconds.

I second is about the length of time it takes to say 'I second'!

Amelia has been playing for 35 seconds.

b) The red hand counts round the clock in seconds.

I know that each mark on the clock face shows one second.

Lee could measure seconds using the clock on the wall, by counting the marks as the second hand moves.

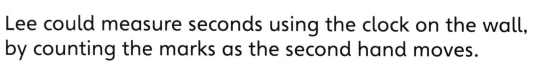

Think together

1 How long does each activity take?

a) Star jumps

Start time   End time

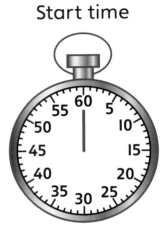 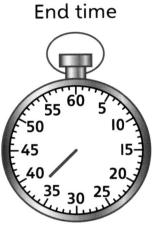 

Star jumps take ☐ seconds.

b) Running

Start time   End time

☐ – ☐ = ☐

Running takes ☐ seconds.

2 48 seconds have gone by.

How many seconds are left until a minute has passed?

I minute = 60 seconds	
48 seconds	?

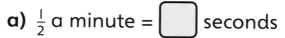 – ☐ = ☐

There are ☐ seconds left.

3 How would you write each time in seconds?

a) $\frac{1}{2}$ a minute = ☐ seconds

b) $1\frac{1}{2}$ minutes = ☐ seconds

c) $2\frac{1}{2}$ minutes = ☐ seconds

I am going to use what I know about the number of seconds in a whole minute to help me.

91

→ Practice book 3C p65

End of unit check

1 One of these months has 30 days. Which is it?

A January

B February

C March

D April

2 Which one of these statements is not true?

A There are 12 months in a year.

B A day lasts from the time you get up to the time you go to bed.

C A day lasts 24 hours.

D A year usually lasts 365 days. Sometimes it lasts 366 days.

3 This clock shows an afternoon time. What time is it?

A 4:58 am

B 2 minutes to 5 am

C 16:58

D 4:55 pm

4 It is 22:47. The duration of a TV programme is 33 minutes. What time will it finish?

A twenty past 11 at night

B 22:80

C twenty past 10 at night

D 11:20 am

5 Molly is timing $1\frac{1}{2}$ minutes. How many seconds is this?

A 60 seconds

B 90 seconds

C $60\frac{1}{2}$ seconds

D $1\frac{1}{2}$ seconds

6 A train starts its journey at this time in the morning.

It finishes its journey at this time in the afternoon.

How long is the train journey?

93

→ Practice book 3C p68

Unit 12
Angles and properties of shapes

In this unit we will …

- ⚡ Learn about turns
- ⚡ Learn what a right angle is
- ⚡ Understand and draw parallel and perpendicular lines
- ⚡ Identify and draw vertical and horizontal lines
- ⚡ Recognise and describe right angles and parallel and perpendicular lines in 2D shapes
- ⚡ Recognise, describe and construct 3D shapes

We will see some different 2D shapes. Which of these are quadrilaterals?

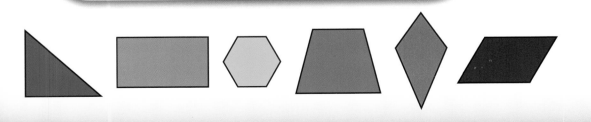

We will need some maths words.
Which of these have you heard before?

right angle   **acute**   **obtuse**   **parallel**
perpendicular   **vertical**   **horizontal**
triangle   **quadrilateral**   **kite**   **trapezium**
rhombus   **parallelogram**   **cuboid**
triangular prism   **square-based pyramid**
cone   **cylinder**   **sphere**   **edges**
faces   **vertices**   **clockwise**   **anticlockwise**

We will look at 3D shapes too. Can you
match the names to all these shapes?

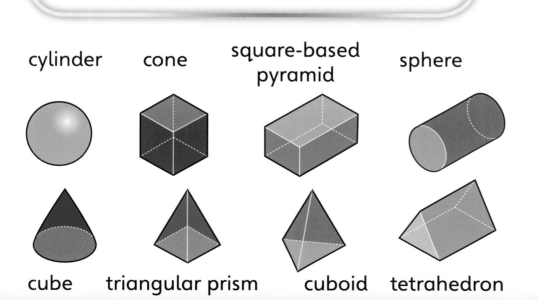

cylinder   cone   square-based pyramid   sphere

cube   triangular prism   cuboid   tetrahedron

Turns and angles

Discover

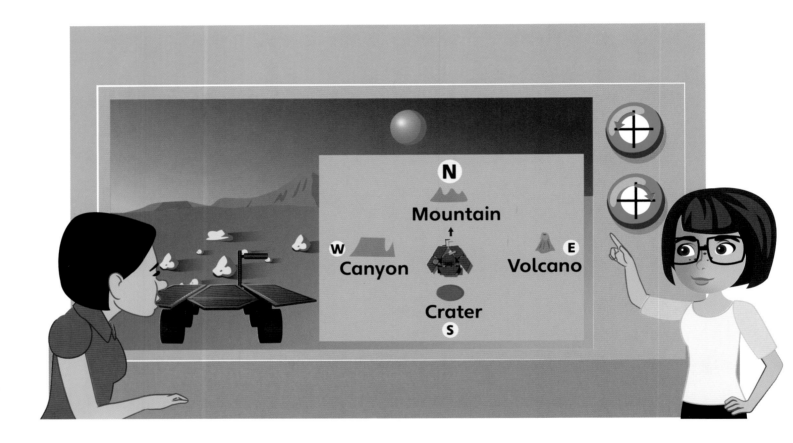

1 **a)** The scientists want to photograph the crater. What instructions should they send to the rover to turn it to face the crater?

b) How would the rover make a complete turn?

Share

a) The instruction buttons are for quarter turns.

> A quarter turn is called a **right angle**. These are both right-angle turns.

anticlockwise　　　　clockwise

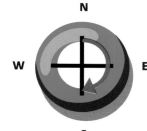

> Two quarter turns make a half turn.

anticlockwise　　　　clockwise

To face the crater, the rover needs to make two quarter turns. It can turn either clockwise or anticlockwise. The final direction will be the same for a half turn.

b) Four quarter turns make a complete turn.

To make a full turn, the rover should make four quarter turns in the same direction. It could turn clockwise or anticlockwise.

> This reminds me of adding fractions. $\frac{1}{4} + \frac{1}{4} = \frac{2}{4}$ which is equal to $\frac{1}{2}$.
> $\frac{1}{4} + \frac{1}{4} + \frac{1}{4} + \frac{1}{4} = 1$

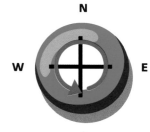

anticlockwise　　　　clockwise

Think together

1 The rover is facing the crater. It makes a quarter turn. Where could it be facing?

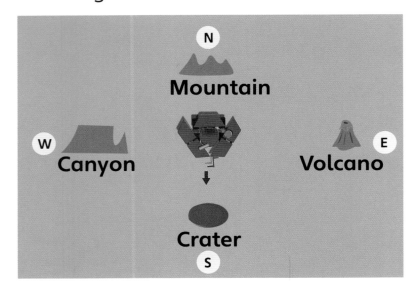

The rover could be facing the _____ .

2 **a)** Now the rover is facing the volcano. It makes three quarter turns clockwise. What is it facing?

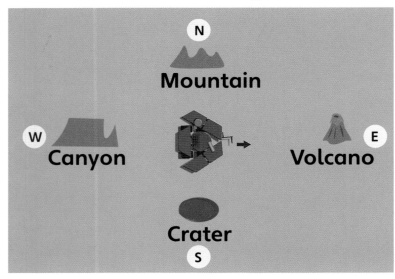

b) From where the rover is now, try three quarter turns anticlockwise. What do you notice?

3

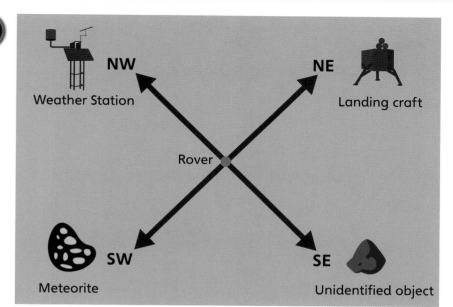

a) The rover turns two right angles. Now it is facing the meteorite. Where was it facing to begin with?

NW means north west.
NE means north east.
SE means south east.
SW means south west.

b) The rover is facing the meteorite. Then it turns to face the unidentified object. Write two different instructions for the turn.

Turning the page around would help!

99

→ **Practice book 3C p71**

Right angles in shapes

Discover

1 **a)** How many of these shapes have at least one right angle?

b) Copy the shapes with right angles onto squared paper and mark each right angle. Show a friend two lines that are perpendicular to each other.

Share

a)

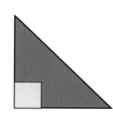

You can check a right angle by making a measurer.

A right angle is shown by a small square in the angle.

Five of the shapes have at least one right angle.

b)

Two lines that are at right angles are **perpendicular** to each other.

Perpendicular lines

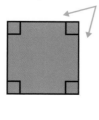   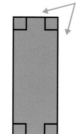   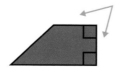   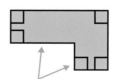

4 right angles   2 right angles   I right angle   5 right angles

Think together

1 How many right angles are on this hockey pitch?

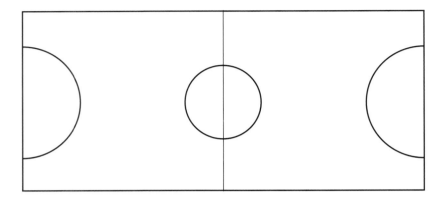

There are ⬜ right angles on the hockey pitch.

This has some curved lines. I do not know about angles in curves.

We only measure angles between straight lines.

2 How many right angles does each shape have?

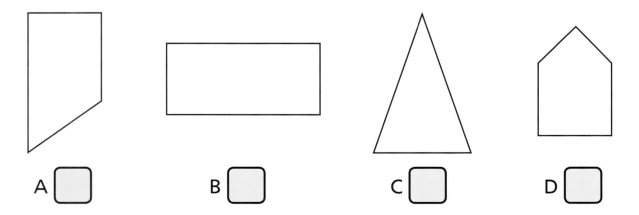

A ⬜   B ⬜   C ⬜   D ⬜

3 Help Dexter to decide if all the angles in each shape are right angles.

I cannot decide if one of these angles is a right angle or a three-quarter turn.

4 How could you draw a line along the dots to make a right angle with each line? Show a partner where you would draw your line.

CHALLENGE

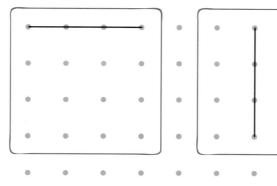

To complete the drawings, I need to draw a line that is perpendicular to the line provided.

→ **Practice book 3C p74**

Comparing angles

Discover

The best roof for snowy or rainy countries has an angle less than a right angle at its peak.

1 **a)** Which house would be good in a snowy country?

b) For countries with little rain, the angle at the peak of the roof is usually greater than a right angle. Do any of these houses suit a dry country?

Share

a)

I need to check if the angles are greater or less than a right angle.

I will use an angle measurer or think about turns.

 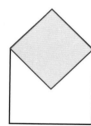 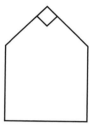   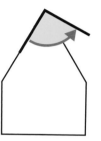

Roof A is a right angle.

Roof C is less than a right angle.

Roof C is less than a right angle. House C would be good in a snowy country.

b) These roofs have angles greater than a right angle.

  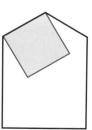 

Roof B is greater than a right angle.

Roof D is greater than a right angle.

Houses B and D would suit a dry country.

Think together

1 Find out if each angle is greater than, equal to or less than a right angle.

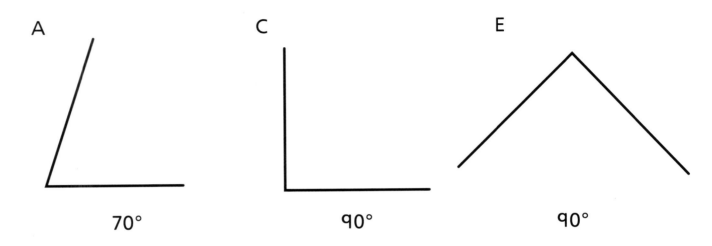

A
70°

C
90°

E
90°

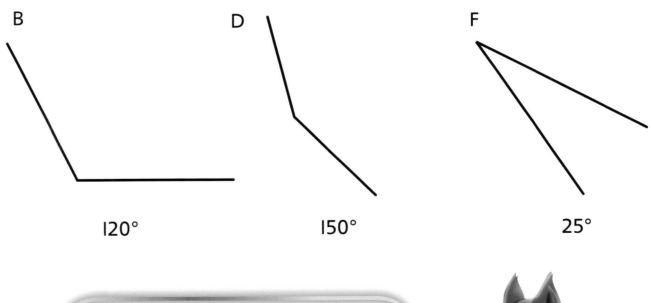

B
120°

D
150°

F
25°

In maths, **acute** means less than a right angle, and **obtuse** means greater than a right angle.

2 Compare each angle with a right angle.

A
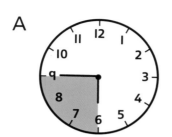

C
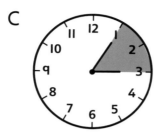

E
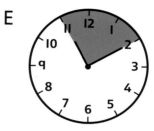

B
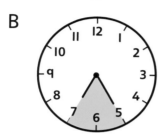

D
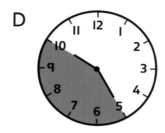

F
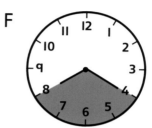

I know the angle between 12 and 3 is a right angle. I wonder which other pairs of numbers make a right angle.

3 Isla has made some angles. She wants to make some different acute and obtuse angles. How many different acute and obtuse angles can Isla make?

CHALLENGE

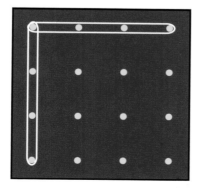

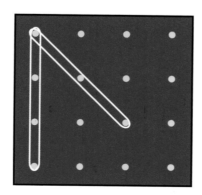

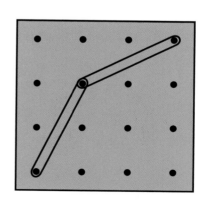

107

Drawing accurately

Discover

To make a 10 cm square I need to measure 10 cm across the top of the paper with my ruler.

I have done the measuring, so now I can cut out my 10 cm squares.

Kate

Richard

1 **a)** How many 10 cm squares can you make out of a strip of paper like Kate's? It is 29 cm and 7 mm wide. How can you do this accurately?

b) How wide is the piece of paper that is left over?

Share

a) Measure 10 cm along the top and along the bottom. Mark 10 cm exactly.

Line a ruler up to **both** marks. Place the pencil on one of the marks to help line the ruler up.

Draw a line to join the two marks. Then cut carefully along the line.

Repeat with the piece that remains.

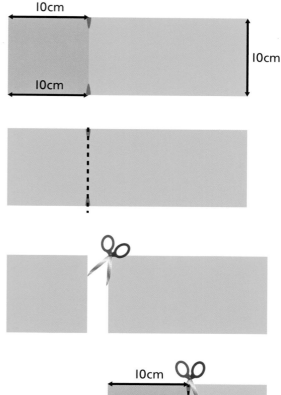

You can make two 10 cm squares. Use a ruler to measure 10 cm accurately across the top and bottom of the strip, make markers, draw a line and cut out the square.

b) The whole strip was 29 cm and 7 mm wide.

20 cm have been cut away.

29 cm – 20 cm = 9 cm

The piece of paper left over is 9 cm and 7 mm wide.

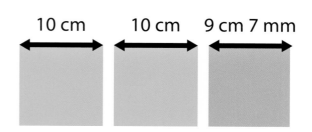

10 cm   10 cm   9 cm 7 mm

Think together

1 Draw these lines on one of the 10 cm squares you have made.

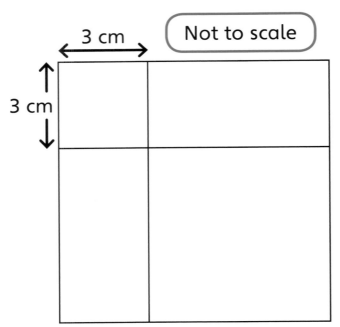

3 cm

Not to scale

3 cm

I could use this method to draw accurate squares of different sizes.

2 Cut along the lines and measure the length of each side.

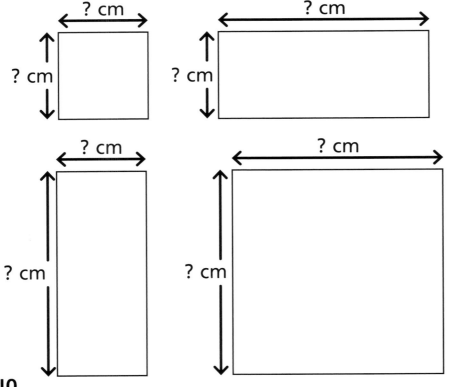

? cm

? cm

? cm

? cm

? cm

? cm

? cm

? cm

I think I can predict the lengths, but I will measure to check how accurate I have been.

Not to scale

3 Draw these shapes using a ruler and a pencil.
Use squared paper to help you.

A

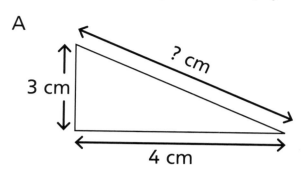

B

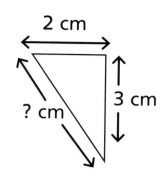

Measure the length of the third side of each triangle accurately.

Triangle A: ⬜ cm   Triangle B: ⬜ cm

4 Aki has a 3 cm square and a 7 cm square.

He draws lines joining opposite corners to make
diagonal lines. Draw these squares and predict the
length of the diagonal lines and then measure to check.

CHALLENGE

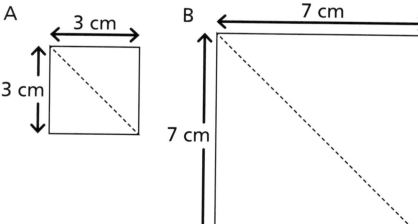

I will try to be
accurate in my
measurements by
using centimetres
and millimetres.

I predict the lines will be 3 cm and
7 cm long because all the sides of
a square are the same length.

III

Types of line ❶

Discover

❶ **a)** Explain why the books have fallen over on one shelf and not on the other.

b) How can the shelf be fixed?

Share

a)

> I know why! The shelf is not straight.
> One end is lower than the other end.

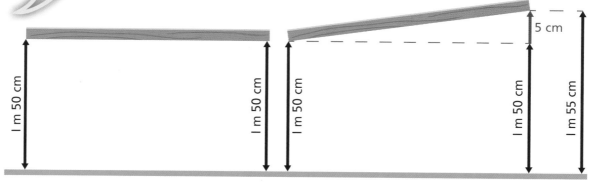

5 cm

1 m 50 cm

1 m 50 cm

1 m 50 cm

1 m 50 cm

1 m 55 cm

One shelf is horizontal. It is the same height all the way along.

One shelf is not horizontal. It is higher at one end than at the other end.

> The word **horizontal** means a straight line that is perfectly level, from left to right.

The books stand upright on the horizontal shelf but have fallen over on the other shelf.

b) The shelf can be fixed by raising one end to 1 m 55 cm, or lowering the other to 1 m 50 cm.

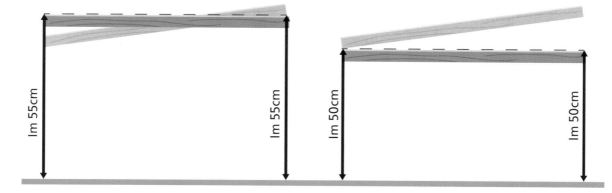

1m 55cm

1m 55cm

1m 50cm

1m 50cm

Think together

1 What is the same and what is different about these fences?

A  B 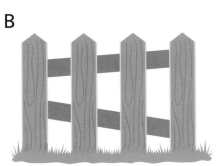 C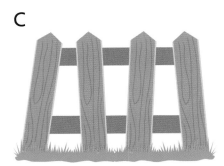

A **vertical** line forms a right angle with a horizontal line.

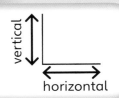

I wonder if I could use the words **horizontal** and **vertical** to answer the question.

2 You can use a plumb line to test if something is vertical. Find some vertical lines to test in your classroom.

A plumb line is a piece of string with a weight on the end. The weight keeps the string hanging straight down, or vertical.

3 **a)** Do any of these shapes have a line of symmetry?
If so, is it vertical, horizontal or both?

CHALLENGE

A

C

E

B

D

F

I remember that lines of symmetry are vertical lines.

I think some of these have horizontal lines of symmetry.

b) Design your own shapes with a horizontal line of symmetry.

115

→ Practice book 3C p83

Types of line ❷

Discover

Isla

Max

1 **a)** Isla folds her paper into a concertina. What will it look like when she unfolds it?

    b) Max has folded his concertina in half. What will his look like when he opens it?

Share

a)

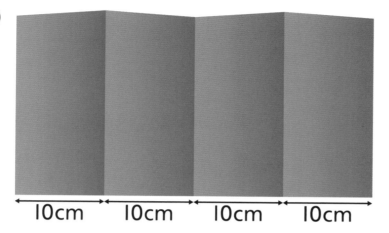

Lines that stay the same distance apart are called **parallel** lines. Lines that cross or meet at a right angle are perpendicular.

When Isla unfolds the paper, she will see parallel lines made by the folds.

b)

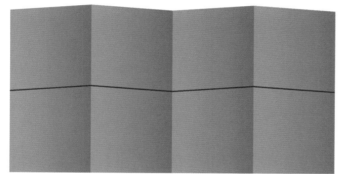

When Max opens his paper, it will have parallel lines like Isla's, but also perpendicular lines where he has folded his paper in half.

I wonder if two parallel lines will ever cross over each other.

Even if you had parallel lines 100 or a million miles long, they would still never touch.

Think together

1 Find the parallel and perpendicular lines in this picture.

Perpendicular lines are at right angles to each other.

I can see both horizontal and vertical parallel lines.

2 Create your own parallel lines by drawing along both sides of a ruler. Why does this create parallel lines?

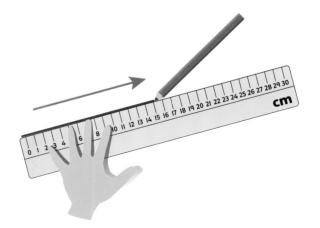

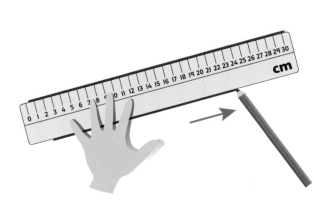

3 **a)** Look at these pairs of lines. Emma says they are
all parallel because they do not cross over.
Explain whether you agree or not.

A   B   C   D

      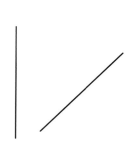

None of these pairs of lines cross over,
but I am not sure if they are parallel.

b) How could you draw lines parallel to each of these?

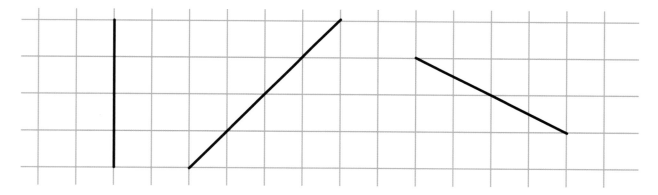

I can draw more than two lines
parallel to the vertical line.

119

Recognising and describing 2D shapes

Discover

1 **a)** What type of quadrilateral have the children made?

How many pairs of parallel lines does this shape have?

b) Do other sizes of this shape have a different number of parallel lines?

Share

a)

I will use sticks to make the shape.

A quadrilateral is a shape that has four sides.

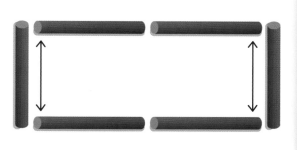

There are two pairs of parallel lines in this rectangle.

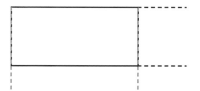

The children have made a rectangle. It has two pairs of parallel sides.

b) All rectangles, including squares, have two pairs of parallel lines.

Opposite sides of a rectangle are the same length. The lines joining them must be parallel.

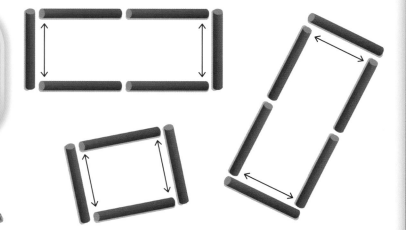

Think together

1 How many pairs of perpendicular lines have the children made?

I will draw some other rectangles on squared paper, to check if the answer is always the same.

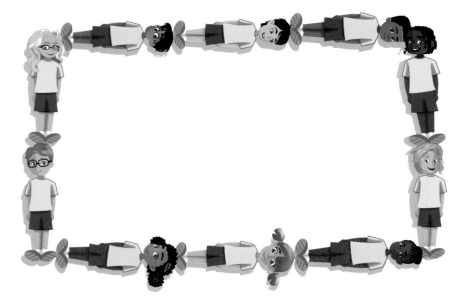

The children have made ☐ pairs of perpendicular lines.

2 **a)** There are six children in the group. Which quadrilateral shapes can they make? Draw them on squared paper.

What about a kite, which has two pairs of equal sides that are adjacent (next to each other)?

I wonder which four-sided shapes they could make with seven people.

b) Which of these shapes has no parallel or perpendicular lines?

3 a) Make some shapes with two acute angles. Use ten sticks or pencils that are all the same length.

> I will make a shape that is not a quadrilateral, like a pentagon.

b) Decide where each of your shapes would go in this table.

	Sides all the same length	Sides not all the same length
No lines of symmetry		
One or more lines of symmetry		

123

→ Practice book 3C p89

Recognising and describing 3D shapes

Discover

Ambika

I think my present is a cube.

Zac

25 cm
22 cm
20 cm

1 a) How can Ambika find out if her gift is a cube?

b) Ambika then describes the faces of her gift. Is it a cube?

Share

a) A cube is a special type of cuboid where all the edges are the same length, and each face is a square.

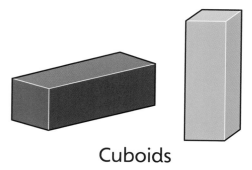

Cuboids

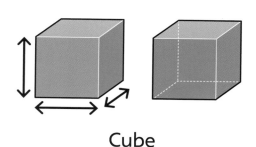

Cube

Ambika can measure all the sides of her gift to find out if it is a cube.

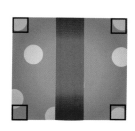

25 cm

22 cm

20 cm

b)

I checked the angles to make sure that each face is a rectangle.

Top / Bottom

22 cm

20 cm

Front / Back

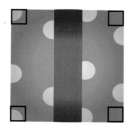

25 cm

22 cm

Side / Side

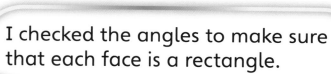

25 cm

20 cm

All the faces of Ambika's gift are rectangles. The opposite faces are exactly the same shape and size. Ambika's gift is a cuboid.

Think together

1 Describe the faces of Zac's gift. How long are the sides of each one?

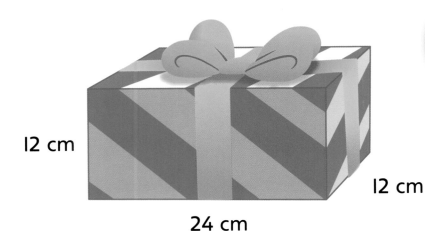

12 cm

12 cm

24 cm

I thought only cubes had square faces.

2 What is the shape of this tent? Count the number of faces, vertices and edges.

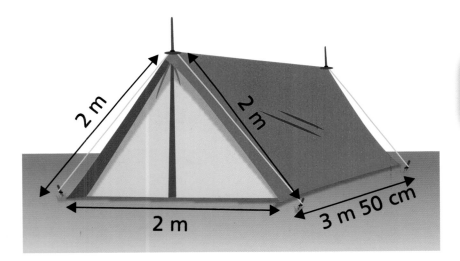

2 m

2 m

2 m

3 m 50 cm

I can describe the measurements of each face.

The shape of the tent is a _____ .

There are ⬜ faces, ⬜ vertices and ⬜ edges.

3 Sort these shapes using the sorting circles.

CHALLENGE

If a shape does not fit into the circles, it must go outside the circles but inside the rectangle.

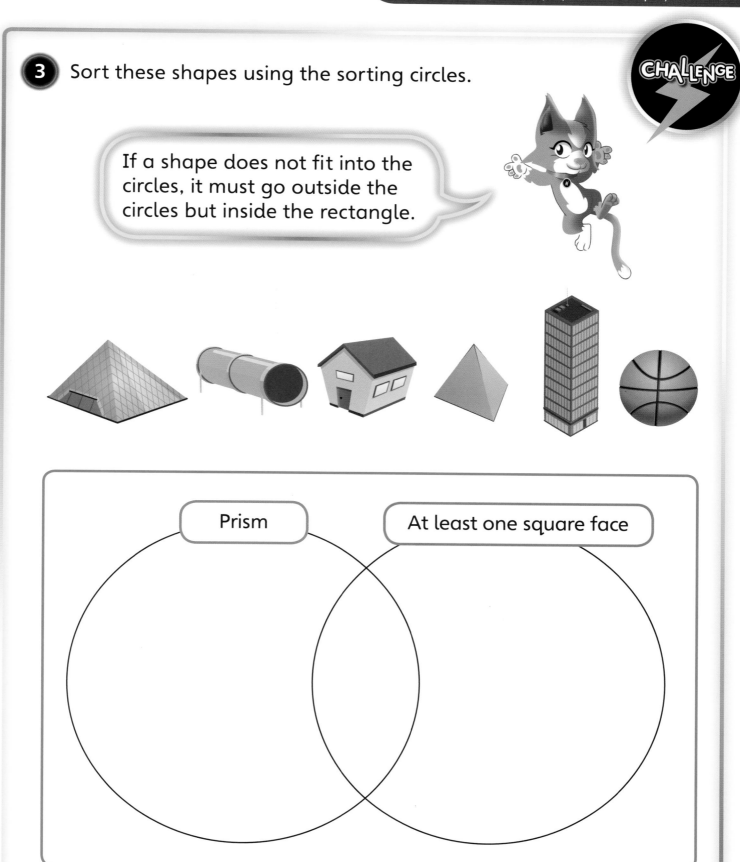

Prism

At least one square face

127

Constructing 3D shapes

Discover

1 **a)** Who can make a cube by putting together their smaller cubes?

b) Lee thinks he can make five different cuboids using all of his small cubes. Is he correct?

Share

a)

Bella's

Lee's

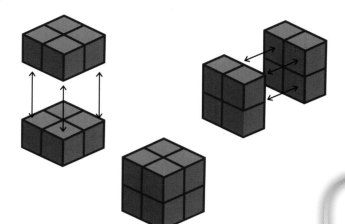

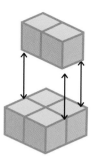

> Lee does not have enough small cubes to make a large cube.

Bella can make a cube by using all of her smaller cubes. It has a length of 2 units in every direction.

b)

Cuboid 2

Cuboid 1

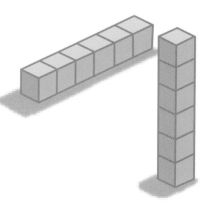

Lee is not correct: he can only make two different cuboids.

He might think he can make five because he can show them from different angles by placing different faces on the table.

Think together

1 Make some cubes using different construction materials.
How many of each will you need?

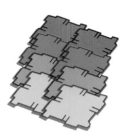  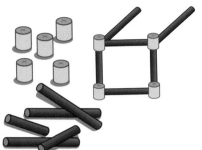 

You need ⬜ squares to make the faces of a cube.

You need ⬜ sticks to make the edges of a cube.

You need ⬜ marshmallows to make the corners (vertices) of a cube.

2 Some children want to make a cube, a sphere, and a pyramid.

a) Which of these shapes can they make from each set of materials?

b) Which shape cannot be made from any of these materials?
Why is this?

| A | B | C |

3 **a)** Which sets of materials could be used to make these 3D shapes?

A  B

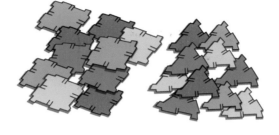

sticks and marshmallows

squares and triangles

linking cubes

I will make a list of which materials can and cannot make each different 3D shape.

I do not think it is possible to make both shapes and I cannot work out how to use the cubes.

b) How many of each piece would you need to make the prism?

131

End of unit check

1 Which shape has just one right angle?

A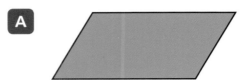

C

B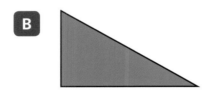

D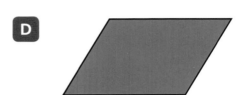

2 Which turn shows an obtuse angle?

A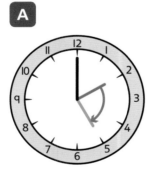

B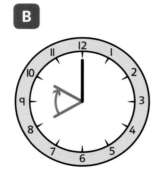

C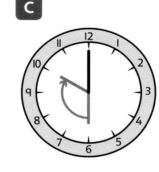

D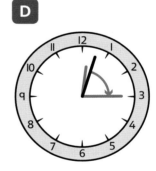

3 Which of these shows three vertical lines?

A

B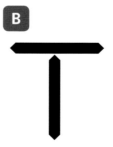

C

D

4 Which shape has no line of symmetry?

A

C

B

D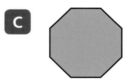

5 Identify the shape with no pairs of parallel lines.

A

C

B

D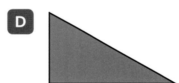

6 Sketch the faces of this shape on squared paper.
Include the measurements.

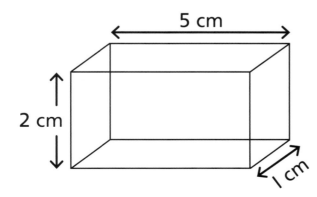

5 cm

2 cm

1 cm

133

→ Practice book 3C p98

Unit 13
Mass

In this unit we will ...
- ⚡ Measure mass in kilograms and grams
- ⚡ Work out different intervals on a scale
- ⚡ Add, subtract and compare masses
- ⚡ Solve problems involving mass

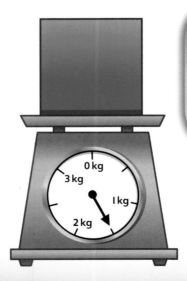

Do you remember what this is called? Use it to find the mass of an object.

We will need some maths words.
Which of these have you met before?

mass   **weigh**   **measure**

scale   **interval**   **grams (g)**

kilograms (kg)

We need to use this too! Use it to work
out the missing number.

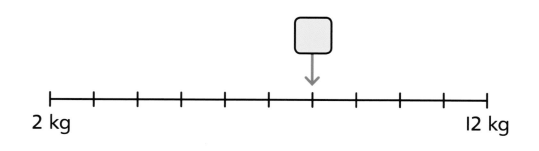

2 kg   12 kg

Measuring mass ①

Discover

① Luis is measuring some clay for a sculpture.

a) How can you work out what each interval represents?

b) What is the mass of each piece of clay?

Share

a)

I used a number line to work out which numbers are missing from each scale.

Each interval is 5 g.

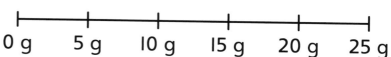

0 g   5 g   10 g   15 g   20 g   25 g

Each interval is 10 g.

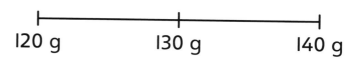

120 g   130 g   140 g

Each interval is 50 g.

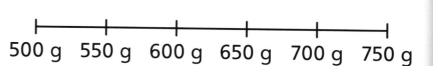

500 g   550 g   600 g   650 g   700 g   750 g

b) The mass of the first piece of clay is 20 g, the second is 130 g and the third is 600 g.

I know that g stands for grams, which is a unit of measure for mass.

Think together

1 Luis weighs more pieces of clay for his sculpture. What is the mass of each piece of clay?

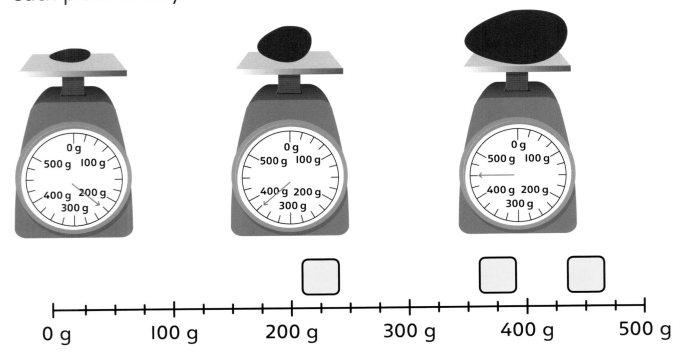

2 Now Luis wants to make a vase and a statue. Decide where the pointer would be on these number lines.

a) Vase: 750 g

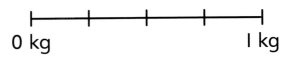

b) Statue: 7 kg

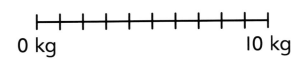

Remember that there are 1,000 g in 1 kg, so half a kilogram is 500 g.

138

a) Sort these objects into the table:

a table, a pen, a T-shirt, a bicycle, a ring, a spoon, a mobile phone, a suitcase

Objects you would measure in grams	Objects you would measure in kilograms

b) A fork weighs 25 g. Use this to estimate the mass of each of the objects listed in your table.

3 kg   120 g   25 g   6 g   10 kg   140 g   6 g   50 kg

139

→ Practice book 3C p101

Measuring mass ❷

Discover

1 a) What is the mass of the bag of carrots?

b) What is the mass of the pumpkin?

Share

a) These scales count in 200s.

The arrow points to between 1 kg 200 g and 1 kg 400 g.

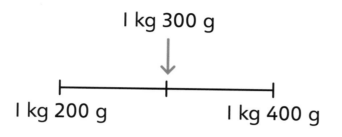

1 kg 300 g

1 kg 200 g   1 kg 400 g

Remember, we always write kg, the bigger unit, before g, the smaller unit.

Half-way between 200 and 400 is 300, so the bag of carrots has a mass of 1 kg 300 g.

b) 7 kg – 6 kg = 1 kg

The difference between the marked amounts is 1 kg. 1 kg = 1,000 g.

There are 10 intervals between 6 kg and 7 kg.

I will try to work out the intervals by counting in 10s, 20s or 100s. I can use a number line to help me!

1,000 divided by 10 is 100.
Each interval = 100 g.

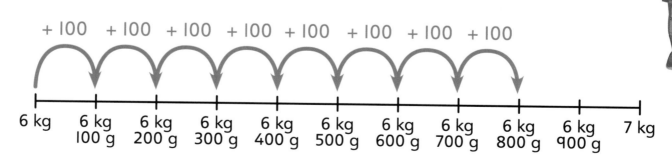

+ 100   + 100   + 100   + 100   + 100   + 100   + 100   + 100

6 kg   6 kg 100 g   6 kg 200 g   6 kg 300 g   6 kg 400 g   6 kg 500 g   6 kg 600 g   6 kg 700 g   6 kg 800 g   6 kg 900 g   7 kg

The pumpkin has a mass of 6 kg 800 g.

Think together

1 Write the mass of these vegetables in kilograms and grams.

Try to work out each interval first by counting on.

a) The onions have a mass

of ☐ kg ☐ g.

b) The peas have a mass

of ☐ g.

2 a) What is the mass of the bag of potatoes?

b) What do you notice about the scales?

3 Use the scales to find the mass of each bag.

a)

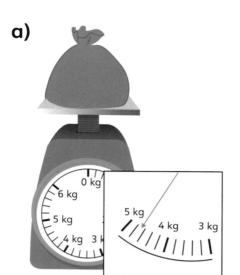

c)

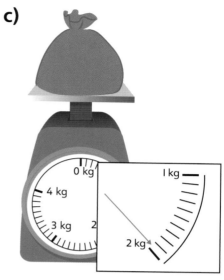

b)

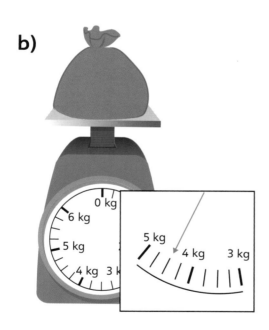

d)

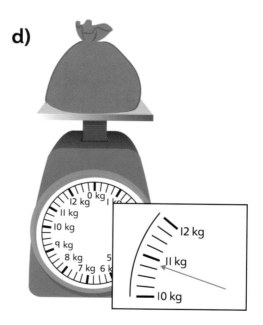

I found the value for each interval and then worked out the midpoint.

143

Measuring Mass ❸

Discover

1. **a)** What is the mass of the bike in kilograms and grams? Whose guess is correct?

   b) Point on the number line below to show the correct mass.

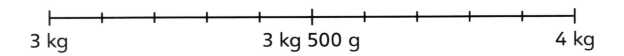

Share

a)

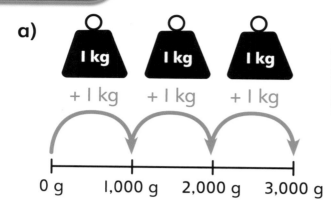

> I counted in 1,000s because I knew there are 1,000 g in 1 kg. I will convert the amount from grams to kilograms and grams.

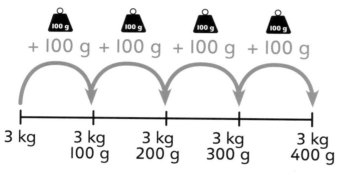

The bike has a mass of 3 kg 400 g.

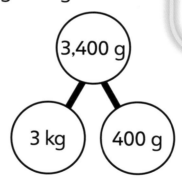

> The part-whole model helps us separate into kilograms and grams.

3,000 g = 3 kg   3,400g = 3 kg 400 g

Bella's guess was correct.

b)

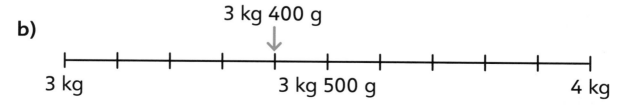

Think together

1 Below is the results table from some more competitions. Work out the missing information.

Result in grams	Result in kilograms and grams
a) 1,232 g	1 kg 232 g
b) 1,567 g	
c)	2 kg 432 g
d)	4 kg 1 g
e) 943 g	

I will not forget to use 0 as a place holder for 4 kg 1 g.

2 Draw the 1 kg and 100 g weights you would use to balance these masses.

a) 6 kg 700 g

b) 2 kg 500 g

I think there is more than one way to balance these masses.

3 Match the equal values.

5 kg 643 g

1 kg 100 g

1,003 g

1 kg 3 g

1,100 g

1 kg 30 g

1,030 g

5,643 g

The part-whole model might help here.

147

Comparing masses

Discover

PINEAPPLES 1,243 g

PUMPKINS 1 kg 230 g

MELONS

1 **a)** Does the pumpkin or the pineapple have the greater mass?
How do you know?

b) The melon weighs more than the pumpkin but less than the
pineapple. How much could the melon weigh?

Share

a) The pineapple weighs 1,243 g, or 1 kg 243 g.

The pumpkin weighs 1 kg 230 g, or 1,230 g.

	Kilograms	Grams
Pineapple	1	243
Pumpkin	1	230

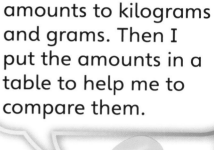

1 kilogram = 1,000 grams

I converted both the amounts to kilograms and grams. Then I put the amounts in a table to help me to compare them.

First, look at the kilograms. These are the same.

Then look at the grams.

243 g > 230 g

The pineapple has the greater mass because 1,243 g is more than 1,230 g.

b) The melon has a mass of more than 1 kg 230 g but less than 1 kg 243 g. The melon could weigh any of the amounts shown below:

I used a number line and found 12 possible answers.

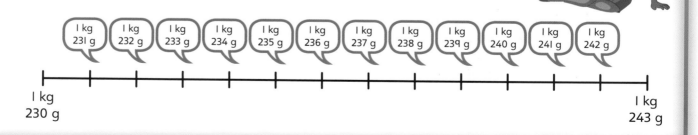

149

Think together

1 Max weighs some more items from the shop. Compare the amounts using <, > and =.

1 kg 456 g ◯ 1,500 g

1,211 g ◯ 1 kg 215 g

1,090 g ◯ 1 kg 9 g

2 kg 211 g ◯ 2 kg 210 g

2 Which scales are not working correctly?

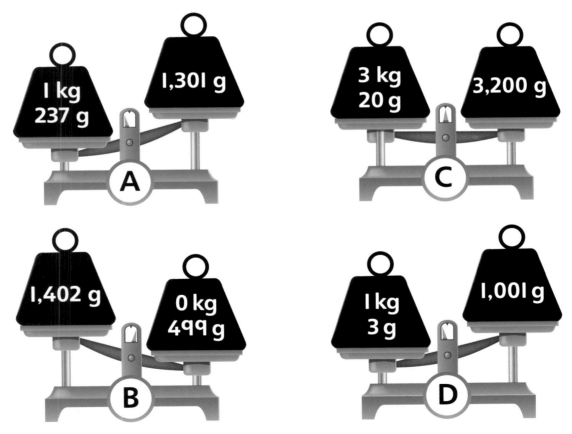

Scales ☐ are not working correctly.

3 Order these masses from lightest to heaviest.

1,432 g   754 g   1 kg 91 g   1,098 g

1,090 g   1 kg 900 g   1 kg 9 g

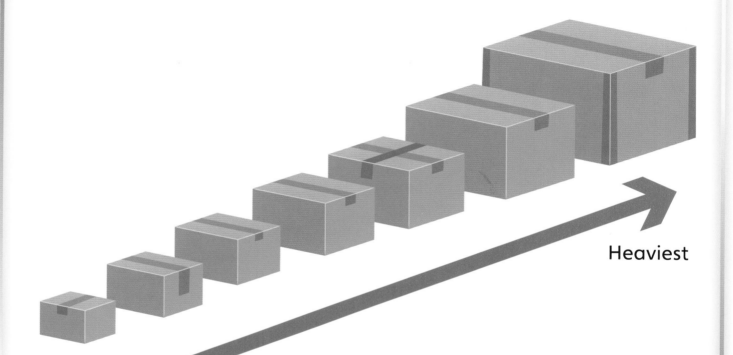

Heaviest

Lightest

I wonder if using a table would help me to compare the masses.

151

→ Practice book 3C p110

Adding and subtracting masses

Discover

I will buy 1 kg 500 g of flour.

I will buy 2 kg 250 g of flour.

Zac

Alex

1 **a)** How much flour do Zac and Alex buy altogether?

b) How much more flour does Alex buy than Zac?

Share

a) Zac buys 1 kg 500 g of flour. Alex buys 2 kg 250 g of flour.
To find the total, use addition:

1 kg 500 g + 2 kg 250 g

First, add the kilograms:

1 kg + 2 kg = 3 kg

Then add the grams:

I used the column method to add the grams.

```
  H  T  O
  5  0  0
+ 2  5  0
  ──────
  7  5  0
```

500 g + 250 g = 750 g

1 kg 500 g + 2 kg 250 g = 3 kg 750 g

Zac and Alex buy 3 kg 750 g of flour altogether.

b) Find the difference between the amounts.

I will use a number line to help me work out the difference.

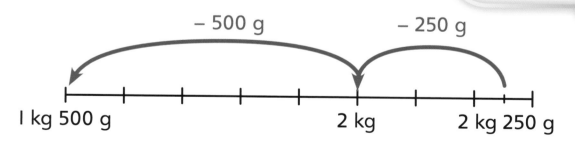

– 500 g   – 250 g

1 kg 500 g   2 kg   2 kg 250 g

Count back 750 g from 2 kg 250 g to get to 1 kg 500 g.

2 kg 250 g – 1 kg 500 g = 750 g

Alex buys 750 g more flour than Zac.

Think together

1 Alex and Zac weigh some more ingredients.
Work out the total mass of each pair.

a) 2,423 g + 1 kg 221 g = ⬜

b) 2 kg 800 g + 200 g = ⬜

c) 1,950 g + 5 kg 100 g = ⬜

I will convert each mass to kilograms and grams. Then I will decide which method to use.

2 Work out the missing mass for each number line.

I had to make more than one jump to find each answer.

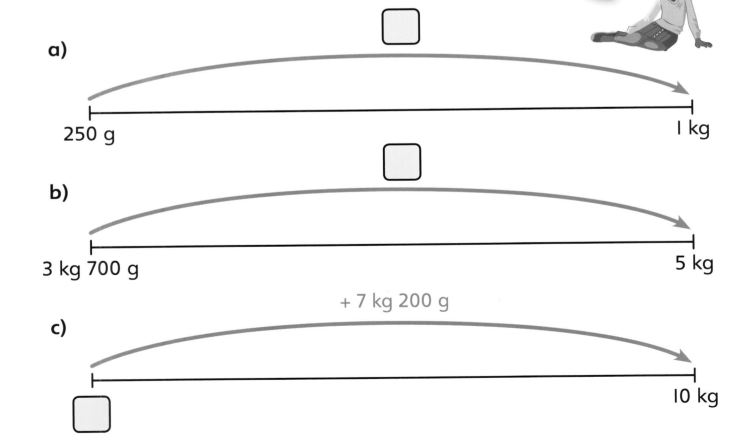

a)

250 g

1 kg

b)

3 kg 700 g

5 kg

+ 7 kg 200 g

c)

10 kg

3 Find the missing numbers in these problems.

CHALLENGE

a)

2,250 g

I kg 500 g	←——————→
	?

I kg 500 g + ⬜ = 2,250 g

b)

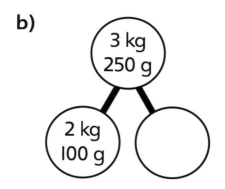

⬜ + 2 kg 100 g = 3 kg 250 g

c)

I kg 900 g	I kg 900 g

?

⬜ − 1,900 g = I kg 900 g

I used the column method to work out answers to some of the questions and a number line for others.

d)

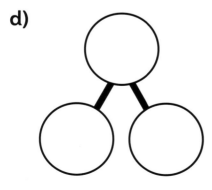

3 kg 500 g − ⬜ = 2 kg 600 g

155

Problem solving – mass

Discover

1 **a)** How much flour does Zac have left?

b) Zac spills 125 g. How much flour does Zac have left now?

Share

a) Zac bought 3 bags of flour. Each bag is 500 g.

500 g × 3 = 1,500 g

Zac bought 1,500 g of flour.

Zac used 1,300 g of flour.

To find how much is left, use subtraction:

1 kg 500 g	
1 kg 300 g	?

1 kg 500 g – 1 kg 300 g = 200 g

Zac has 200 g of flour left.

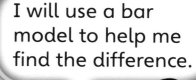

I will use a bar model to help me find the difference.

b) To find out how much is left, use subtraction:

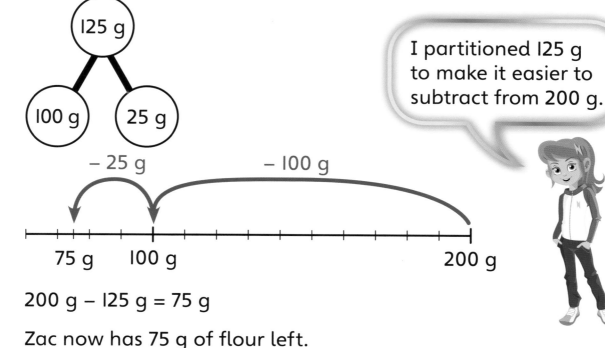

I partitioned 125 g to make it easier to subtract from 200 g.

200 g – 125 g = 75 g

Zac now has 75 g of flour left.

Think together

1 Zac weighs some flour on the scales. He adds 50 g of butter and 275 g of sugar. The scale now shows the total mass.

What is the mass of the flour?

The mass of the flour is ⬚ g.

2 2 sweets with a mass of 15 g each and 3 chocolate bars with a mass of 40 g each are put on the scale.

What mass will the pointer be pointing to?

I will find the total of the 2 sweets and 3 chocolate bars first, by adding.

The pointer will be pointing to _____ .

3 What is the mass of 1 wagon?

CHALLENGE

I wonder if I should find the mass of 1 robot first.

159

→ Practice book 3C p116

End of unit check

1 How much does the box weigh?

A 250 g

B 220 g

C 350 g

D 250 kg

2 Which is equal to 1,040 g?

A

C

B

D

3 Which mass is greater than 1 kg 255 g?

A 1 kg 99 g   B 256 g   C 1,275 g   D 1 kg 100 g

4 Which calculation gives the answer I kg 350 g?

A 100 g + 250 g

C 1,900 g – 1 kg 650 g

B 900 g + 550 g

D 1 kg 800 g – 450 g

5 What is the total mass of the boxes?

A I kg 340 g

B 1,350 g

C I kg 200 g

D 140 g

6 A child's bike weighs 1,256 g. Her friend's bike weighs 300 g less. How much do they weigh altogether?

161

→ Practice book 3C p119

Unit 14
Capacity

In this unit we will ...

⚡ Measure capacity in litres and millilitres

⚡ Convert between litres and millilitres

⚡ Compare and order capacities

⚡ Add and subtract capacities

⚡ Solve problems involving capacities

Do you remember using a bar model to add numbers? Use this one to find the total.

350	500
?	

We will need some maths words.
Which ones have you seen before?

capacity   **litre (l)**   **millilitre (ml)**

scale   **interval**   **convert**

Can you use part-whole
models to partition numbers?

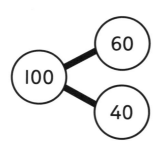

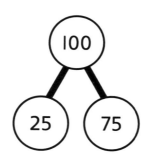

Measuring capacity ①

Discover

① **a)** The zookeeper has to make sure that the elephant has enough water to drink. How much water is left in the tank?

b) The parrot's water bottle holds 200 ml. How many millilitres is each marker on the scale? How many millilitres of water are in the bottle?

Share

a)

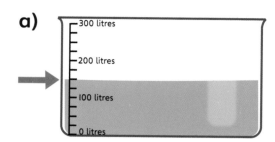

The water level is half-way between 100 litres and 200 litres. I need to work out which number is exactly half-way.

I could use a number line to find the intervals between the markers.

We measure capacity in millilitres (ml) and litres (l). Capacity is the total amount a container can hold.

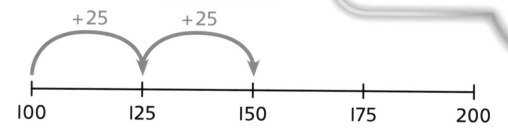

150 litres is half-way between 100 litres and 200 litres. There are 150 litres of water left in the tank.

b) Each 100 ml is divided into 5 sections. 100 ml ÷ 5 = 20 ml. Each marker is 20 ml.

The water level is 100 ml + 20 ml + 20 ml = 140 ml
There are 140 ml of water in the bottle.

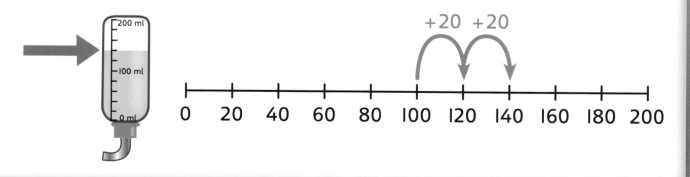

Think together

1 How much fuel is left in the tank?

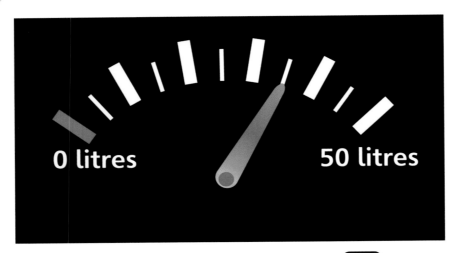

The scale is marked in intervals of ☐ l.

$7 \times$ ☐ $=$ ☐ l

There are ☐ l of fuel left in the tank.

2 A recipe needs 275 ml of milk. Where is 275 ml on the jug's scale?

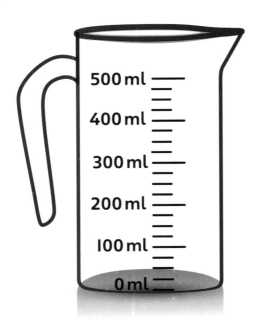

I need to work out what each mark between the numbers is worth.

3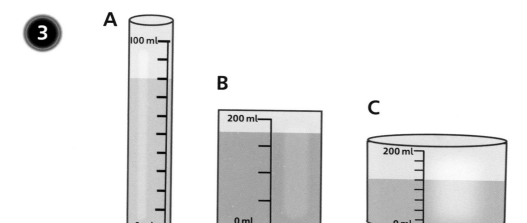

A

100 ml

B

200 ml

0 ml

C

200 ml

0 ml

0 ml

a) How much liquid is in each container?

b) Put the containers in order from least amount of liquid to most amount of liquid.

It looks like A has the most liquid because it is higher.

All the containers have different scales. I need to check each one carefully.

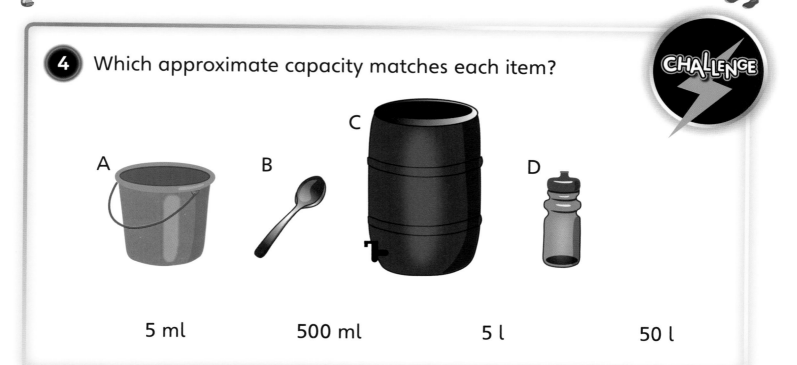

4 Which approximate capacity matches each item?

CHALLENGE

A

B

C

D

5 ml

500 ml

5 l

50 l

167

Measuring capacity ②

Discover

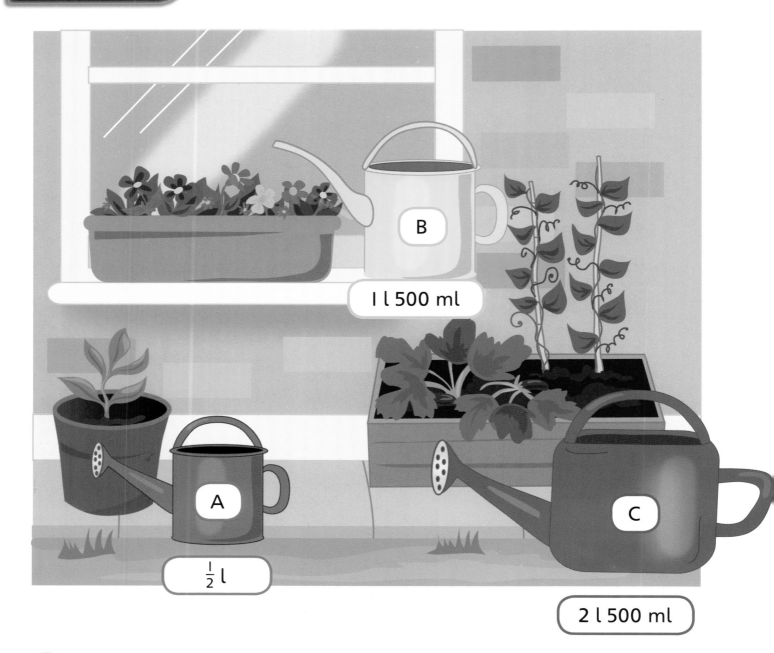

B

I l 500 ml

A

$\frac{1}{2}$ l

C

2 l 500 ml

1 **a)** How many millilitres does each watering can hold?

b) Write the capacity of watering can B in whole and half litres.

Share

a)

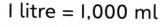
I litre = 1,000 ml

A: 1 l = 1,000 ml

$\frac{1}{2}$ l = 1,000 ml divided by 2 = 500 ml

$\frac{1}{2}$ l = 500 ml

Watering can A holds 500 ml.

B: 1 l 500 ml = 1 l + 500 ml

= 1,000 ml + 500 ml = 1,500 ml

Watering can B holds 1,500 ml.

I started by partitioning into litres and millilitres.

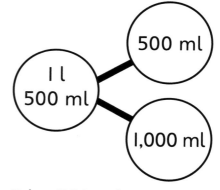

C: 2 l 500 ml = 2 l + 500 ml

2 l = 1,000 ml + 1,000 ml = 2,000 ml

2,000 ml + 500 ml = 2,500 ml

Watering can C holds 2,500 ml.

b) 1 l 500 ml = 1,000 ml + 500 ml

= 1 litre + $\frac{1}{2}$ litre = 1 $\frac{1}{2}$ litres

1 l 500 ml	
1 l	500 ml
1 l	$\frac{1}{2}$ l
1 $\frac{1}{2}$ l	

Think together

1   How much milk is in the jug?

I know it is half-way between 1 l and 2 l. How do I write that?

2 l (= 2,000 ml)

I could use a number line to help me.

←

200 ml = 1 l 200 ml

100 ml = 1 l 100 ml

1 l (= 1,000 ml)

There is ⬜ whole litre and ⬜ ml in the jug.

There are ⬜ ml in the jug.

2 **a)** How much liquid is in each container?

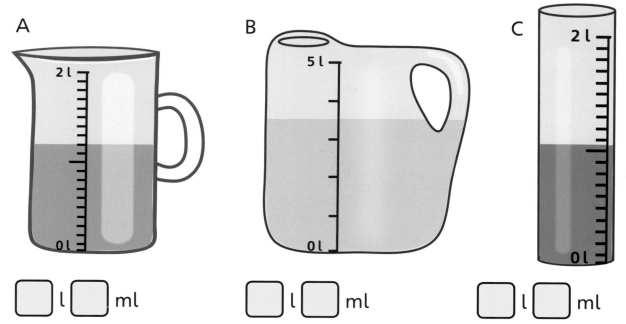

A ⬜ l ⬜ ml   B ⬜ l ⬜ ml   C ⬜ l ⬜ ml

b) Write each amount in ml.

A = ⬜ ml   B = ⬜ ml   C = ⬜ ml

3 Which scale shows 1 l 400 ml? Explain how you know.

CHALLENGE

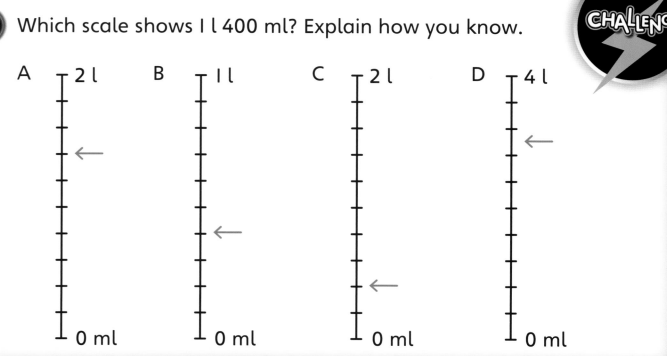

171

Measuring capacity ③

Discover

Choux Pastry

440 g butter
600 g flour
1,200 ml water
16 eggs

① **a)** The recipe asks for 1,200 ml of water. Where will that be on the scale?

b) The baker uses 3 l 250 ml of cream for the filling. How many millilitres is this?

172

Share

a)

1,200 ml	
1,000 ml	200 ml
1 l	200 ml

There are ten intervals. Each interval is 100 ml.

1,200 ml = 1 l and 200 ml on the scale.

2 l

+100 ml

+100 ml

1 l 200 ml

1 l

I will use bar models to answer this question.

b)

3 l 250 ml			
1 l	1 l	1 l	250 ml
1,000 ml	1,000 ml	1,000 ml	250 ml
3,250 ml			

3 l 250 ml is 3,250 ml of cream.

Think together

Fruit Punch:

Ingredients

1 l 600 ml orange juice

2 l 750 ml apple juice

4,250 ml lemonade

1 These are the ingredients for a fruit drink.

I will try using bar models or part-whole models for these questions.

a) How many millilitres of orange juice are needed?

1 l 600 ml = ☐ ml of orange juice.

b) How many ml of apple juice are needed?

2 l 750 ml = ☐ ml of apple juice.

2 l 750 ml		
1 l	1 l	☐ ml
☐ ml	☐ ml	☐ ml
☐ ml		

c) What is 4,250 ml of lemonade in l and ml?

4,250 ml = ☐ l ☐ ml

4,250 ml				
1,000 ml	1,000 ml	☐ ml	☐ ml	250 ml
☐ l	☐ l	☐ l	☐ l	250ml
☐ l				250ml

2 The baker needs a bowl big enough to hold 2 l 350 ml of water.

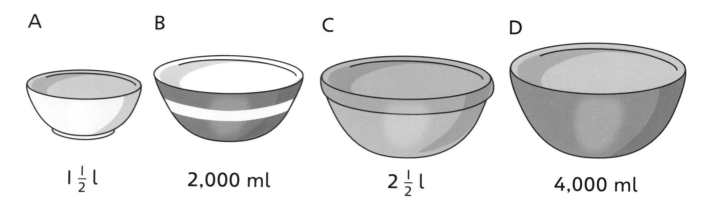

A $1\frac{1}{2}$ l

B 2,000 ml

C $2\frac{1}{2}$ l

D 4,000 ml

Which bowl should he choose?

3 The baker puts 1 l of milk and 50 ml of cream into the jug. How many millilitres of liquid will he have altogether?

CHALLENGE

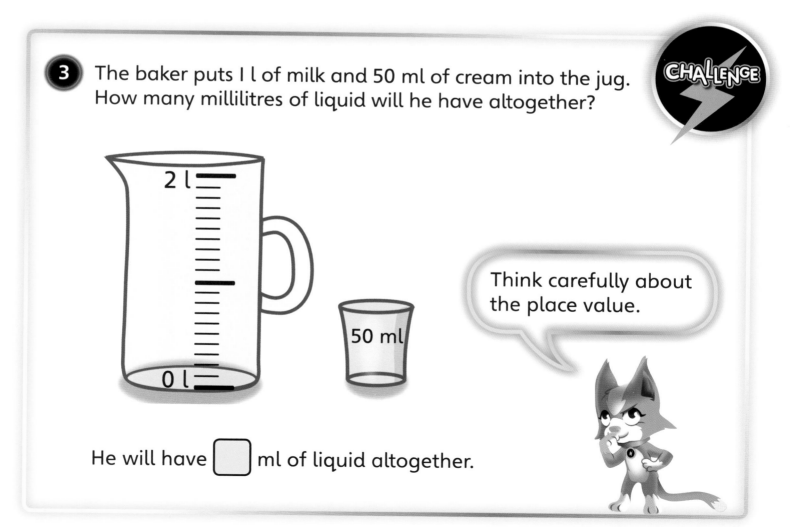

Think carefully about the place value.

He will have ☐ ml of liquid altogether.

175

Comparing capacities

Discover

orange juice
2 l 400 ml

water
2,500 ml

lemon squash
1 l 450 ml

apple juice
$1\frac{1}{2}$ l

1 **a)** Is there more orange juice or more lemon squash?
Show how you know.

b) Put the drinks in order of amount, from the most to the least.

Share

a)

I will try looking at the number of litres in each.

orange juice
2 l 400 ml

lemon squash
1 l 450 ml

The orange juice has **2** litres, the lemon squash has only **1** litre. There is more orange juice.

b)

I need to think about both the litres and the millilitres.

orange juice

2 l	400 ml

water

2,000 ml	500 ml

lemon squash

1 litre	450 ml

apple juice

1 litre	$\frac{1}{2}$ l
1 litre	500 ml

2,500 ml > 2 l 400 ml

There is more water than orange juice.

$1\frac{1}{2}$ l > 1 l 450 ml

There is more apple juice than lemon squash.

The drinks in order of amount from most to least are: water, orange juice, apple juice, lemon squash.

Think together

I will look at the millilitres because both cars have I litre. But which one has more ml?

1 a) Which toy car has the greater fuel capacity?

A 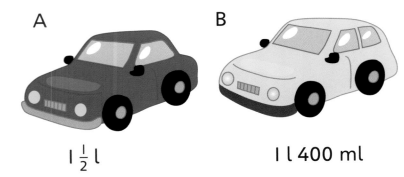 B

$1\frac{1}{2}$ l   I l 400 ml

$\frac{1}{2}$ l = ⬜ ml.   Car _____ has the greater capacity.

b) Which is less, 2 l 900 ml or 2,850 ml? Explain how you know.

c) Which is greater, $3\frac{1}{2}$ l or 3,550 ml? Explain how you know.

2 Put the containers in order, from smallest to greatest capacity.

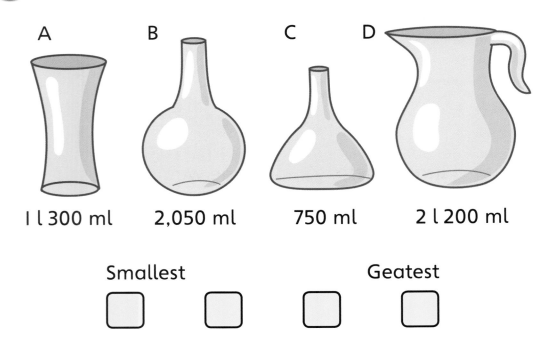

A   B   C   D

I l 300 ml   2,050 ml   750 ml   2 l 200 ml

I will look at the litres first.

Smallest   Geatest

⬜ ⬜ ⬜ ⬜

3 Reena and Max are buying fizzy drinks.

Reena buys a multi-pack of eight cans. Each can contains $\frac{1}{2}$ l.

Max buys two bottles. Each bottle contains 2 l.

Use the number line to work out who has more fizzy drink.

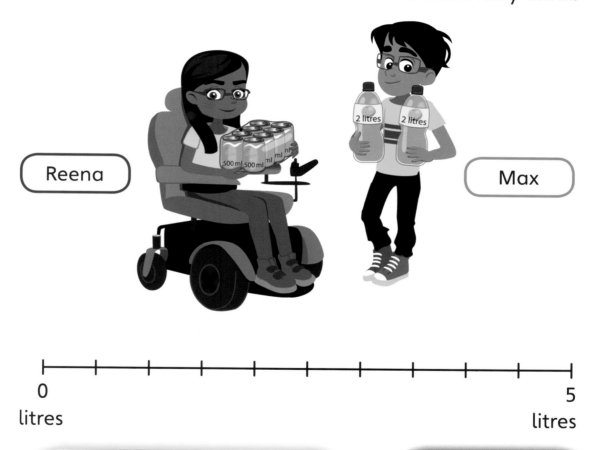

Reena

Max

0
litres

5
litres

I think I can use
500 millilitres = $\frac{1}{2}$ litre to help.

I think a number
line would help.

179

→ **Practice book 3C p130**

Adding and subtracting capacities

Discover

Doctor Crawford

Professor Smith

1 **a)** Which scientist has less than 1 l of liquid altogether?

b) How much more liquid does one scientist have than the other?

Share

I will use 1 l = 1,000 ml and 500 ml = $\frac{1}{2}$ l to help me.

a)

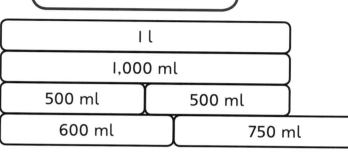

Doctor Crawford

| 1 l |
| 1,000 ml |
| 500 ml | 500 ml |
| 600 ml | 750 ml |

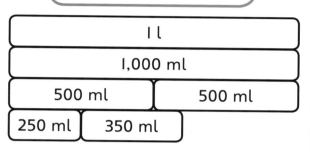

Professor Smith

| 1 l |
| 1,000 ml |
| 500 ml | 500 ml |
| 250 ml | 350 ml |

Doctor Crawford: 750 ml and 600 ml are both greater than 500 ml. So 750 ml + 600 ml must be greater than 1,000 ml

Professor Smith: 250 ml and 350 ml are both less than 500 ml. So 250 ml + 350 ml must be less than 1,000 ml

Professor Smith has less than 1 litre of liquid altogether.

b) Professor Smith is using 600 ml. Doctor Crawford is using 1,350 ml.

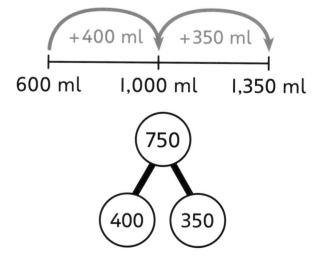

I used a number line to help me find the difference.

400 ml + 350 ml = 750 ml

The doctor has 750 ml more liquid than the professor.

Think together

1 How much more liquid will you need to fill the jugs to 1 litre?

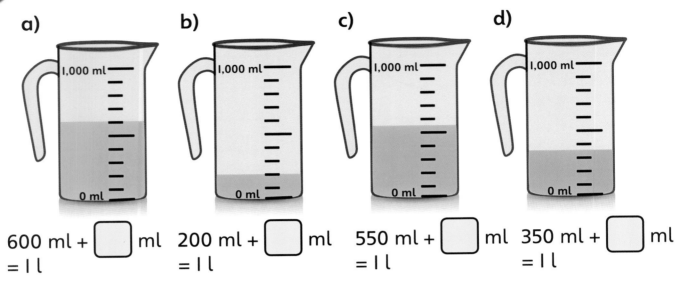

a)

600 ml + ☐ ml
= 1 l

b)

200 ml + ☐ ml
= 1 l

c)

550 ml + ☐ ml
= 1 l

d)

350 ml + ☐ ml
= 1 l

2 A litre jug is filled with water. 250 ml is poured out into a glass. How much is left in the jug?

☐ ml is left in the jug.

3 What is the total capacity of these jugs?

1 l 400 ml

2 l 250 ml

I think I can add the litres and millilitres separately.

1 l + 2 l = ☐ l

400 ml + 250 ml = ☐ ml

Total capacity = ☐ l ☐ ml

```
  H  T  O
  4  0  0
+ 2  5  0
_____
```

4 Both drinks are poured into one 3 l jug. How much more liquid is needed to fill the 3 l jug?

CHALLENGE

800 ml

1 l 400 ml

First of all, I will add the litres and millilitres.

1 l	400 ml	800 ml
1 l	☐ ml	

☐ l ☐ ml

How many ml are there in 3 l? That should be easy as I know that there are 1,000 ml in 1 l.

3 l	
☐ l ☐ ml	☐ ml

☐ ml more liquid is needed to fill the 3 l jug.

→ Practice book 3C p133

Problem solving – capacity

Discover

Pancakes
30 g butter
300 g plain flour
600 ml milk
3 eggs

Holly

1 **a)** How many cartons of milk does Holly use for the pancake recipe?

b) She pours equal amounts of the juice into five cups. How much juice will there be in each cup?

Share

a) The recipe says 600 ml milk.

Each carton contains 200 ml.

> I will keep adding 200 ml until I reach 600 ml.

Pancakes

30 g butter
300 g plain flour
600 ml milk
3 eggs

600 ml		
200 ml	200 ml	200 ml

Holly uses three 200 ml cartons of milk for the pancake recipe.

b) The bottle contains 1 l of juice.

> How do I divide 1 l into 5 equal parts?

I litre

> I will divide 1,000 ml by 5 instead. I know that 10 divided by 5 is 2.

1,000 ml				
200 ml	200 ml	200 ml	200 ml	200 ml

There will be 200 ml of juice in each cup.

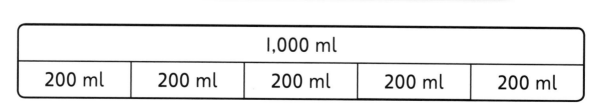

Think together

1 **a)** Francesca uses 180 ml of shower gel in 6 days.
How much does she use each day?

180 ml					
⬜ ml	⬜ ml	⬜ ml	⬜ ml	⬜ ml	⬜ ml

Francesca uses ⬜ ml each day.

b) Does Francesca have enough shower gel left for another day?

200 ml – ⬜ ml = ⬜ ml. Is that enough for one day? _____

2 Francesca drinks 2 l 250 ml of water one day and 1,250 ml the next day. How much water does she drink altogether in the two days?

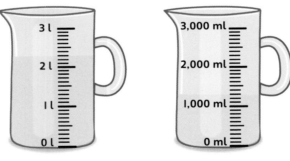

I will change the millilitres into litres and millilitres, then add the litres and millilitres separately.

2 l 250 ml	1,250 ml
2 l 250 ml	⬜ l ⬜ ml
⬜ l ⬜ ml	

Francesca drinks ⬜ l ⬜ ml altogether.

3 Leon has a 2 l bottle of water and some cups.
Each cup holds 250 ml.

CHALLENGE

2 l

250 ml

Remember, 1 l is 1,000 ml.

a) How many cups can he fill from the bottle?

Leon can fill ☐ cups.

b) If Leon fills six cups, how much water is left in the bottle?

6 cups = ☐ ml.

There are ☐ ml of water left in the bottle.

I could use a number line or a bar model to help me.

187

End of unit check

1 Which cylinder contains 350 ml?

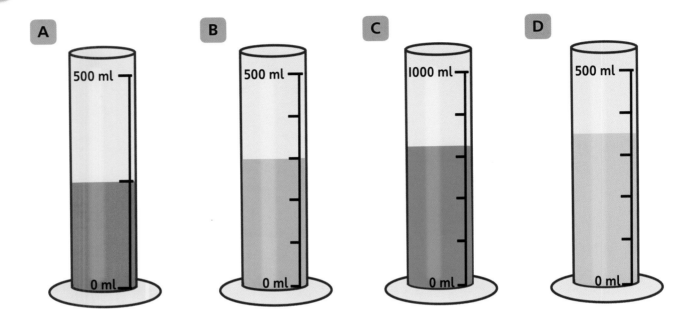

2 Which jug shows 1 l 200 ml?

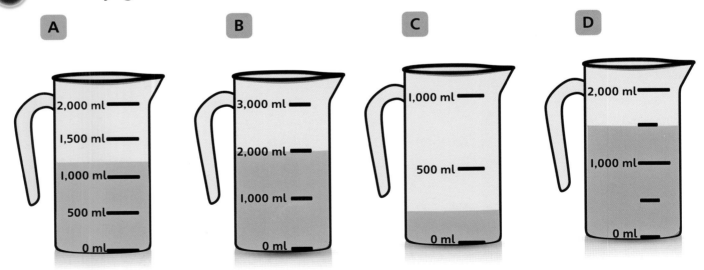

3 What is 2 l 50 ml in millilitres?

A 1,050 ml   B 2,050 ml   C 2,500 ml   D 250 ml

4 Which is the largest capacity?

A 3 l 200 ml   B 3 l 50 ml   C 2 l 800 ml   D 2 l 90 ml

5 What is 3 l 600 ml + 5 l 700 ml?

A 8 l 300 ml   B 9 l 200 ml   C 9 l 300 ml   D 8 l 1,300 ml

6 What is 5 l 200 ml – 3 l 400 ml?

A 2 l 800 ml   B 1 l 800 ml   C 1 l 900 ml   D 2 l 200 ml

7 A bottle holds 1 l of lemonade.

Rachel fills 5 glasses with lemonade.

She puts 150 ml into each glass.

How much lemonade is left in the bottle?

→ Practice book 3C p139

Holiday fun

Here are some ideas you can try at home.

Rainy Day!

Invent your own board game that uses scores up to 1,000. Use your knowledge of addition and subtraction with numbers up to 1,000 to create a game where the aim is to score 1,000 points before your partner.

There could be bonus squares or challenge cards where you win or lose 10, 20, 100 or even 500 points!

Make sure the rules are fair and easy enough for a new player to learn.

Exercise games

Make up an exercise routine that includes moves such as running, jumping, skipping or even moving through an obstacle course. Use the game to practice times-tables, by chanting different tables for each exercise. For example, you might do 10 star jumps and count in the five times-table as you go!

An extra challenge could be to count down from 36 in the three times-table.

Be inventive and challenge yourself.

Measuring

Create an accurate map of your house or garden. Measure the lengths and widths carefully and include as many details as you can in your map.

If you have some help from an adult, you could go for a walk around your local area and count the paces as you go. Use this information to create a map of the area, including distances to important landmarks.

Investigate whether it takes the same number of paces to walk to a place as it does to walk back again. Make sure you have suitable adult supervision.

Baking or cooking

With the help of a suitable adult, use your measuring skills to create a meal, or bake a cake or a loaf of bread. You will need to research suitable recipes and use measuring scales accurately to make sure you have the correct weight or volume of each ingredient.

What we have learnt

⚡ Find equivalent fractions and compare two fractions

⚡ Add and subtract two fractions with the same denominator

⚡ Understand what years and months are

⚡ Tell the time using analogue and digital clocks and use 12-hour and 24-hour clocks

⚡ Find and compare durations of time

⚡ Find start and end times and measure time in seconds

⚡ Identify turns

⚡ Identify right angles in shapes and compare angles

⚡ Identify and find horizontal and vertical symmetry, parallel and perpendicular lines

⚡ Draw and measure straight lines accurately

⚡ Recognise and describe 2D and 3D shapes

⚡ Measure and compare masses and capacities

⚡ Add and subtract masses and capacities

⚡ Convert grams and kilograms, millilitres and litres

Can you do all these things?

I tried everything and learnt lots, especially when I made mistakes.

Now you are ready to continue your maths journey in Year 4.